BACKYARD BEEKEEPING

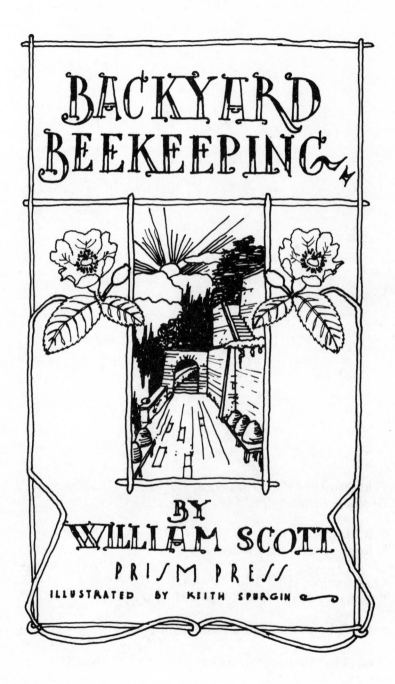

BACKYARD BEEKEEPING

BY WILLIAM SCOTT

PRISM PRESS

ILLUSTRATED BY KEITH SPURGIN

Published 1977 by: Prism Press, Stable Court, Chalmington, Dorchester Dorset. DT2 OHB

Cloth edition: ISBN 0 904 727 43 2
Paperback edition: ISBN 0 904 727 44 0

Set IBM by 𝗧 Tek-Art, Croydon, Surrey.

Printed by: Unwin Brothers Ltd, the Gresham Press, Old Woking, Surrey.

A swarm of bees in May
Be worth a load of hay;
A swarm of bees in June
Be worth a silver spoon;
Swarming in July
Let the buggers fly.

Anon

Acknowledgements

I would like to thank the various people who have helped with the production of this book.

Firstly, my parents, to whom it is dedicated: my father whose encouragement helped turn my desire to 'be a writer' into a reality; my mother, who gave me my first lesson in beekeeping.

Thanks also to Adrian Morris Thomas and Trevor Tolputt, who provided some interesting and unusual bee literature from their own collections; Les, with whom much of my knowledge of bees was gained 'the hard way'; Coral, for bringing her fastidious Virgoan faculties to bear on the manuscript; Brenda, for help with the typing; Shirley for her comments.

Most special thanks to Keith who, as well as producing the delightful drawings that enliven the following pages and providing much detailed information about flowers and their relationship with bees, has been the main source of inspiration and encouragement throughout the writing of this book.

Contents

About this Book

ot so long ago, before we in Britain
became dependent on imported sugar and
various synthetic substances for sweetening our food and drink, a great many
families in the country kept bees.

Today, people grow up with the idea
that beekeeping is a complex, highly
skilled and often extremely dangerous
occupation. Just as we are taught that only artists really know
about painting pictures, only farmers care about the land and
only priests know much about God, so we might easily go
through life imagining that the art of beekeeping was not for
the likes of us.

This little book sets out to show you that anyone, anywhere, can keep bees. It describes the life of the bee, the very
basics of beekeeping, including a section with detailed
instructions on how to make your own equipment, and it
concludes with some notes on the various bee products and
their uses in the home.

You will also find a list of more comprehensive books to
which you may want to refer as you go deeper into the
subject.

In the meantime this volume contains enough information
to whet your appetite and get you started. From there on it
is mainly your own experience and intuition that will trans-
form you into a beemaster.

Why Keep Bees?

hy not? You can do it almost anywhere — there is no need even to own a piece of land: a wise farmer would be happy to let you position your hive amongst his crops: some even pay for such a service!

The basics of beekeeping are easily learnt; the commitment of time is small — even the busiest businessman should have the leisure to carry out the minimum necessary work during the year; the rewards, on the other hand, can be great.

"It is not an occupation to be undertaken by indifferent or indolent people. Energy and perseverence, together with method, power of observation, and aptness to act quickly in an emergency are fundamental attributes for the making of a proficient beekeeper."

So said the veteran beekeeper W. Herrod Hempsall in 1938. True enough. But don't let such strident statements frighten you away from a pastime that is as demanding as you care to make it. After all, colonies of bees survive and prosper year after year without man's assistance. What Mr. Hempsall is talking about is the ability to manipulate the natural events in a hive to your advantage.

I have heard someone refuse a teaspoonful of honey on the grounds that it was produced by exploiting one of God's creatures and, while one does read of heavy-handed practices advocated by some 'experts' — clipping the wings of queen bees, etc — it must be said that a good beekeeper does no more than give direction to the activities of the hive, at the same time as ensuring that the bees are well protected and supplied for the winter months, removing the surplus for himself. There need be no more exploitation involved here than in the case of the gardener who ties his tomato plant to a pole!

The first reason for getting yourself a hive of bees must be the simple fascination of watching nature at work. The 'mysteries' of the honey bee have absorbed men from childhood to old age throughout history. Whether you are intrepidly scaling the higher reaches of a tree in pursuit of a swarm, or quietly sitting nearby listening to the bees' precise imitation of the cosmic hum, you are being absorbed, recreated, developed.

The next obvious reason for keeping bees is one of economics. Getting started can cost you a small fortune — about £50 per hive — or it can cost you nothing at all — read on! For some reason the honey from your own hive tastes unlike any you can buy. With just one strong hive in a good year you will have enough honey to replace the deadly white stuff for all sweetening purposes throughout the year. The great value of honey as a food and remedy, its other uses, and those of the bees' other products will be discussed later. If you intend keeping bees on a larger scale there's money to be made.

There are of course the indirect benefits of keeping a hive: if you have a garden, especially fruit trees, a nearby hive ensures maximum pollination, and therefore more produce. Roughly eighty percent of all pollination in fruit orchards is carried out by the honey bee.

While at certain times of the year there is intensive labour involved on the part of the beekeeper, mainly at the end of the summer when it's time to harvest the honey crop, most of the time hives can be left alone and the bees allowed to get on with it. You don't have to have someone in to look after them when you go away for the weekend.

Bees can be kept almost anywhere, provided there is a bit of shelter and space for them to take off and land. You don't even need to have a garden as bees have been kept successfully on roofs in the centre of London. They will make use of vegetation on other people's property and there are no laws forbidding your bees from taking nectar from someone else's flowers. Bees will fly up to two miles foraging and there can't be many places without sufficient flora within that radius to sustain a hive of bees.

16

Bees can be kept almost anywhere

However, if you are fortunate in having a fair amount of land, the possibility of actually growing suitable crops for your bees is interesting. A colony situated in a field of clover has only a few yards to fly on each trip and, provided conditions are right, honey yields can be enormous (over one hundred lbs per hive).

Before describing the techniques of beekeeping let's first take a look at the life of the colony as it continues from year to year with or without the interference of the beekeeper.

PART ONE:

The Hive and the Honeybee

The Hive

n the wild the bees' home may be found almost anywhere: in a hollow tree, the roof of a house, under a stone, even suspended in the open air amongst the branches of a tree. More familiar to us are the artificial homes or hives, constructed by man to enable him easily to manipulate the colony and obtain large amounts of surplus honey without harming the bees.

Whether a colony of bees is living in the wild or under the care of a human beekeeper it behaves in much the same way. In fact the practice of keeping bees in anything more elaborate than a hollow log or straw basket (skep) is very recent.

Upon occupying suitable premises, the bees will set about constructing comb. This they do with scales of wax secreted from beneath their abdomen. The cavities or cells in the comb, the shape of which is so well known, serve as storage vessels for food, resting places, and spaces in which young bees can be reared. Between all the layers of comb is always found a gap of about eight millimetres known as the 'bee space', by means of which the bees move around the hive.

Great wonder is often expressed at the hexagonal shape of the honeycomb, though we should not be surprised to see nature accomplishing the task in the easiest, most direct way. The size of the cell is decided by the size of the bee, just as humans build houses and furniture to a human scale. If you place a number of circles together and compress them slightly to remove the gaps between them, hexagons are produced.

The bees build two sizes of comb depending upon the size

The Skep, – outside and inside

of the bee that is to grow therein, the larger cells being for the larger drone (male) bees. A third kind of cell, in which the Queen bees are reared, is occasionally constructed. Generally the comb in the lower part of the hive is used for brood rearing (bring up baby bees) while the comb above is used to store honey.

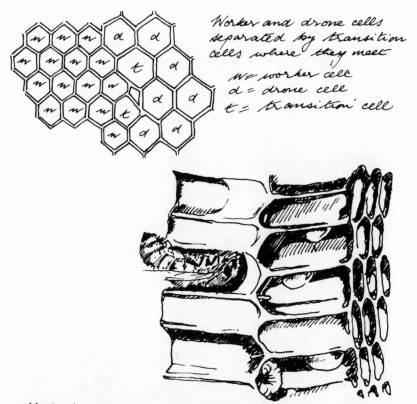

Worker and drone cells separated by transition cells where they meet

w = worker cell
d = drone cell
t = transition cell

Having begun work on the comb, the framework around which their whole life evolves, the bees put the outer structure of the hive in order. Any draughty gaps are sealed, any roughness smoothed over, and if the entrance is too large for safety it will be made smaller. All the structural work is carried out with a resinous substance collected from plants and trees called 'Propolis', from the Latin meaning 'in front of the city', so called because in the wild it is most easily observed around the entrance of the hive.

Once established in its home, a successful colony may remain there indefinitely.

The Inhabitants

Kingdom:	Animal
Phylum:	Arthropoda
Class:	Hexapoda
Order:	Hymenoptera
Family:	Apidae
Genus:	Apis
Species:	Mellifica

he honey bee is found throughout the world and comes in many different shapes and sizes. Widespread are the brown bees, native of Great Britain and most of Europe. They are slow but steady workers, very hardy and resistant to disease and can be easily handled.

There are other dark European bees. The Dutch bee, being bad-tempered and a poor producer, has not gained much popularity. The Carniolan, from around Austria, is a much less common, larger bee, prone to swarming but very economical with propolis and hence good for comb honey production.

Caucasian bees are very gentle but strong mountain bees that do not swarm much and work very hard. It is surprising that this strain is not more widely used: it seems to combine the best qualities of all the varieties. However the best known strain these days is the Italian or Yellow bee. Hard workers, though often hot tempered, they are regular in their swarming habits and resistant to disease. Other less well-known yellow bees are found in the Near East and Cyprus and many other varieties exist throughout the world, each with its own characteristics.

There is for example the giant bee of India that makes comb up to six feet long; the oasis bee of North Africa that will not sting; and the Chinese bee that will work in temperatures only a few degrees above freezing. (Note 1).

Many beekeepers now breed their own strains to suit their own requirements and local environment. A well-known example is Brother Adam of Buckfastleigh Abbey in Devon who has produced a hybrid that is ideal for keeping in the south west of England.

Whatever the strain of bee, the colony will always be made up of a queen bee, a large number of workers and a smaller number of drones.

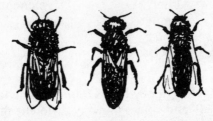

Drone   Queen Worker

Hive bees belong to an extensive order and have many 'cousins'. Some common allies, the wasp, cuckoo-bee and hoverfly, are illustrated.

The Queen

The life of every colony centres around the queen bee; perhaps 'mother' would better describe her status, since her sole function is the procreation of bees, and at no time does she rule the colony but is rather its servant: the hour of her birth and even the hour of her succession is decided by other members of the hive.

The body of a queen is long and tapering, with shortish wings. She is the only fully developed female in the colony and her lifelong task is the production of eggs. This she does with extraordinary efficiency — up to three thousand eggs a day at her peak during the spring and summer months, the time when the maximum number of workers is required. Without the presence of a queen the colony lacks the means to regenerate itself, though provided with a female larva under two days old the workers can, by rapidly constructing a queen cell around it, turn the situation to their advantage and nurture a new queen. Thus it will be seen that both queens and the workers begin life in the same way as 'female' or fertilised eggs. Drones on the other hand, develop from 'male' or unfertilised eggs.

What it is that determines whether a 'female' egg develops into a queen or a worker is not yet fully understood. Obviously the specially constructed cell in which the queen takes form is an important factor. These queen cells are much larger than normal cells and are constructed in a perpendicular fashion with the opening at the bottom. They are about eight millimetres in diameter and twenty-five millimetres long, and are found in the hive only at the time of queen rearing; once the princess has emerged and the period of change within the hive is over the queen cells are usually demolished by the workers.

A Queen Cell

Another influence on the different growth patterns of workers and queens is their diet. A substance, known as Royal Jelly, is fed to all larvae during the first three days of growth, but only the queen larvae continue to receive large quantities of this potent food right up to the time they turn into pupae (the stage of transition from grub to bee). After these first three days the workers and drones are fed a mixture of jelly, pollen and honey, and the quantity is reduced. Obviously there are properties in this rich, milk jelly itself that affect the growth of the larvae, but probably the strongest influence is the amount they are fed. Compared with the plentiful royal diet, that of the workers is after the third day meagre, and their smaller size and lack of sexual organs is probably due to partial starvation (note 2).

The different stages of the metamorphosis of a queen are as follows:- the egg hatches after three days, five and a half days later the larva is full grown and spins its cocoon and remains in the pupal stage for seven and a half days before emerging as a fully developed virgin queen. The whole process therefore, takes about fifteen days. The fact that workers spend almost twice as long in the pupal stage may also have a strong influence on their differing development.

However unclear the reasons for the differences between worker and queen, the differences themselves are obvious. As well as being much larger, the queen alone possesses properly functioning reproductive organs. It is however the worker who, though usually unable to lay eggs, has the brood-food glands which enable her to nurture the young, a faculty not enjoyed by the queen. Neither does the queen possess wax glands, pollen baskets, a suitably shaped tongue or other 'implements' for the collection and storage of supplies. The queen's sting is long and curved without barbs and she uses it only against a rival queen. The worker uses its short barbed sting at the first scent of danger. A 'female' larva reared as a worker is constantly caring for the needs of the queen in the hive; if reared as a queen she will do her best to destroy any other queen she meets. Queens have lived for up to seven years whereas a worker in the honey season may only live for as many weeks.

When the queen is ready to emerge from her cell she bites at the capping whilst slowly turning herself round. Finally, when the circle is almost complete, the end swings open like a door.

She climbs out, eats a little honey and at once sets about establishing her authority. Any other queens that are present in the hive, usually with the exception of her own Mother, are polished off either by stinging or, in the case of undeveloped pupae still in their cells, simply by ripping open the capping.

Within a week the queen has left the hive a few times; at first to get her bearings and then for her nuptial flight. This occurs usually but once and always when the air is warm around midday, this being the time when most drones are likely to be out and about. The nuptial flight lasts between ten and thirty minutes.

The queen leaves the hive and, followed by a number of drones, flies off into the summer afternoon. Very few people have seen the actual mating since it always takes place in the air, but it is known that the queen yields herself only to one drone, presumably the strongest and fittest since he must outfly all the other drones. At the point of contact there is said to be a loud cracking sound, caused by the drone exposing his genital organs which are normally concealed within his body. This process has been likened to the sudden popping that occurs when one tries to turn the fingers of a rubber glove inside out by blowing into them! At this point the drone dies. As soon as contact is made the couple fall to the ground. The queen frees herself and, taking with her a portion of the unfortunate drone's genitals, returns to the hive never to mate again. Consequently, during this one mating she must obtain enough spermatozoa to last her whole egg-laying career. This seems to be the reason that she carries the drone's genitals back to the hive with her, for it can take up to seven hours for all the spermatozoa to penetrate into the queen's spermatheca, a storage sac from which they are released one by one to fertilise eggs when required.

Two or three days later the queen begins laying.

"When the queen is about to lay, she puts her head into a cell and remains in that position for a second or two, to ascertain its fitness for the deposit she is about to make. She then withdraws her head, and curving her body downwards, inserts the lower part of it into the cell; in a few seconds she turns half round upon herself and withdraws, leaving an egg behind her." Rev. L.L. Langstroth.

If a queen is observed in the hive, worker bees will continually be seen to approach and lick her. It seems that she secretes some substance, the smell of which is immediately recognised by all the bees, and that once this chemical is absent from the hive the workers know it is time to start rapidly rearing a new queen.

Only in deepest winter does the queen cease laying and she begins again before spring to build up the hive population in readiness for foraging. As the queen gets older her egg-laying capacity drops, sometimes her supply of spermatozoa runs out, and she lays only unfertilised or 'male' eggs which will produce drones even though laid in worker cells.

The queen under normal conditions, will die of old age after about three or four years.

The Worker

The worker is a female bee, an undeveloped queen, and whereas there is only one queen in the hive there are many thousands of workers.

When laid, the 'female' or fertilised egg is attached to the bottom of a cell by a sticky substance. There it remains for about three days. As soon as it hatches out the bees who are concerned with caring for the young (nurse bees) supply the tiny worm with Royal Jelly. Floating in this rich food the larva rapidly grows, shedding successive skins.

After the sixth day the larva's diet is changed somewhat, pollen and honey being added to the jelly. On the eighth day the cell is sealed over and the cocoon is spun, so that by the tenth day the larva is settled, head facing the opening of the cell, ready for the transformation into the pupal stage. The last skin is shed the next day and thereafter the pupa remains still until twelve days later (the twenty-first day) when the pupal skin is lost and the young bee is ready to emerge. These changes do not occur suddenly, the whole process being continuous and gradual.

"At a rate that is imperceptible, the wormlike body of the larva gradually is reconstructed into three distinct body regions, while the larva's organs become modified and converted into those necessary for adult life. Eyes, antennae and mouth parts appear on the head, wings and legs on the thorax, and the abdominal region of the larva acquires adult form. Colour changes in general become apparent first on the anterior parts of the body and last on the posterior. Pigmentation shows up first in the compound eyes which, on the thirtieth day change to pink, then red, purple, and finally to brown by the time of emergence." (Note 3).

Having once emerged the young bee begins to work. Its first duties are domestic, it cares for the young and the queen, and keeps the hive air conditioned (not too hot, not too cold, and at a constant humidity) by means of fanning the wings. It tidies up, mends the hive with propolis, takes care of most of the wax production and comb building, receives the nectar from the "field" bees arriving in the hive, manipulates it in such a way that it loses some of its excess moisture, and stores it away in the cells.

As it becomes older (after another 21 days) its duties take

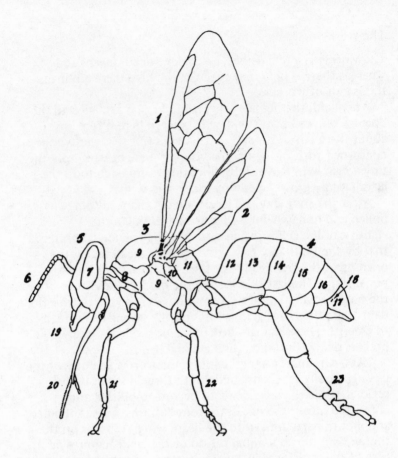

The Worker, external structure

1 & 2. Wings
3. Thorax
4. Abdomen
5. Head
6. Antenna
7. Compound Eye
8. Prothorax
9. Mesothorax

10. Metathorax
11. Propodeum
12-17. Abdominal Segments
18. Spiracle
19. Mandible
20. Proboscis
21-3. Legs.

it outside the hive in search of nectar, pollen, propolis and water.

The main physical characteristics of the worker are all connected with the labours she performs ceaselessly throughout her life. Most interesting are the pollen baskets, a collection of tiny curved hairs on the hind legs, so-called because they are used to hold large bundles of pollen for transportation to the hive, where it is used as a protein-rich food.

Pollen, basket

Sting of Worker

A. left side of complete apparatus
1. attachment of quadrate plate
2. quadrate plate
3. sheath lobe
4. hinge of triangular plate
5. hinge of triangular plate on ablong plate
6. ablang plate
7. triangular plate
8. apex of triangular plate
9. basal arm of lancet
10. basal arm of bulb and stylet
11. bulb of stylet
12. shaft of sting
13. stylet
14. lancet.
B end of lancet, much magnified

The best-known feature of the worker is the short barbed sting located at the tip of the abdomen with which those bees on guard duty will attack any intruder. The use of the sting means almost certain death for the bee since it must leave its sting and part of its abdomen attached to the victim. Along with the sting it seems that a chemical substance is released which communicates to the rest of the hive that there is danger afoot. Bees will only use their sting in the vicinity of the hive; if threatened in the field they will not stay to argue.

Comb building is another task of the young worker bee. The wax for this is secreted in the form of scales by glands underneath the abdomen. These scales are then masticated with saliva and shaped as they are applied to the cells. Only the young house bee can produce wax, and then only when the air is quite warm and they have eaten large amounts of honey.

Nectar collection is achieved by means of a proboscis which is inserted into that part of the flower where the nectar accumulates (the nectary). It is then stored in the bee's honey sac for later transference to the bees in the hive. Water is collected in the same way.

Most flowers produce nectar in minute quantities so that the bee must visit at least 50 blossoms, and sometimes as

Horse Chestnut and visitor

many as 1,000, to fill her honey sac (note 4). The amount of nectar obtained from one blossom will depend upon the type of plant, the temperature and humidity, and whether or not the flower has recently been visited by another insect.

With her honey sac full the worker returns to the hive and distributes her bounty among a number of *'house'* bees whose job is to process and store it. If the nectar source is abundant the worker "dances" to inform other field bees. The "round dance" tells of a source of nectar within one hundred yards of the hive.

"The loaded nectar carrier shakes her abdomen from side to side, all the while running in arcs of circles, turning first to one side and then to the other. Usually she is followed by four or five other bees and while she continues her dance, every now and then, one of the interested followers may be seen to leave for the field until, by the time the dancer is ready to depart, a dozen or more may have left to search out the source of the rich find" (note 3).

There is another more fascinating type of dance that can be observed within the hive. It has been called the "wagtail dance" and the movements actually communicate the direction in which the supply lies. Outside the hive the bee uses the sun as a means of orientation; inside the hive this is described in terms of gravity! When the bee dances down the comb she demonstrates that the source is away from the sun; up the comb means that it is towards the sun. The enthusiasm with which these dances are performed is proportional to the proximity of the blossoms described. The whereabouts of pollen, water and propolis are communicated in the same way.

36

Once the nectar has been gathered by the field bee, the process of changing it into honey begins. While in the honey sac "inversion" starts to occur (the change from cane sugar to simple dextrose and levulose). Then when the house bee takes charge she will manipulate it for about half an hour to allow evaporation of a large proportion of the moisture. This thickened honey is then placed in a cell where further evaporation takes place in the warm dry atmosphere of the hive. After about three days there is a concentration of 80% sugar and the cell is sealed. This honey is then 'ripe'. It is interesting to consider just how much 'bee labour' is needed to produce one pound of such ripe honey. Bear in mind that for every pound of wax that bees produce they must consume up to twenty pounds of honey. In a hive there may be sixty thousand bees at work, but if we imagine *one* bee working alone, continously, day and night, all through the year, it has been estimated that it would spend *eight years* producing one pound and that during that time it would have flown a distance equal to more than two circumferences of the earth! (note 3).

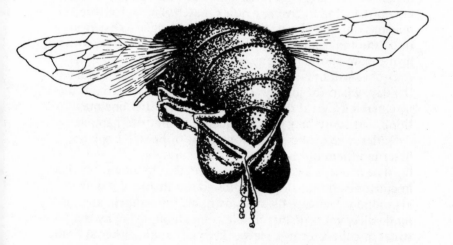

The Drone

The drone is the male of the species. Drones are produced only during the spring and summer months when they may be needed for mating with a queen.

Drone larvae are produced from unfertilised eggs. They are larger than the workers and are reared in special larger cells. Drones have no sting. They collect no nectar, no pollen, produce no wax, and do no work in or out of the hive. This does not mean however that they do not consume their fair share of honey which they beg from the hive workers.

Once any possible need for drones is past and forage is becoming scarcer, the workers withhold food from them, threatening to sting them and finally when they are weak from hunger, drive them from the hive. It is interesting that though there may be hundreds of drones in the hive with a queen it is only when the queen leaves the hive that the drones show any interest in her. On such an occasion any drone in the vicinity will pursue her with great ardour and the strongest and fittest will catch her and mate with her, thus loosing his genital organs and his life, but passing on his superior qualities and characteristics to subsequent generations of bees. It seems that the reason why more than one or two drones are reared in each colony is that when the queen *does* make her marriage flight, it's important that plenty of drones are on the wing to ensure a quick match. The longer the queen is out and about, the greater the danger to her from birds and the elements. Frequently a queen will mate with a drone from another colony.

Drones will only be seen outside the hive in the middle of the day, when the air is warm. Often they hover around the hive making a great deal of noise so that at first one may think that something of consequence is about to happen.

Butler, one of the earliest observers of bees in England, describes them nicely, if not quite accurately:

"The drone is a gross, stingless bee, that spendeth his time in gluttony and idleness. For howsoever he brave it with his round velvet cap, his side gown, his full paunch, and his loud voice, yet is he but an idle companion, living by the sweat of others' brows. He worketh not at all, either at home or abroad, and yet spendeth as much as two labourers; you shall never find his maw without a drop of the purest nectar. In the heat of the day he flyeth abroad, aloft and about, and that with no small noise, as though he would do some great act; but only for his pleasure, and to get him a stomach, and then returns he presently to his cheer." (note 5).

... in the middle of the day when the air is warm'.

Winter. As the temperature drops sunlight weakens and the days become shorter, so the colony (workers and queen) clusters into a quietly vibrating bundle to maintain warmth. This cluster moves slowly around the hive consuming the honey, its only source of energy, so carefully stored away in the days of plenty. This is not hibernation; there is always movement: those on the colder outside of the cluster pass to the centre and vice versa in a continual flow like the convection of air.

As temperatures drop, the cluster becomes tighter, more compact; and with milder weather it expands. On very mild days bees can be seen flying from the hive for the purpose of dumping their faeces. But in the depth of Winter, if you stand beside a healthy hive, you might well suppose it to be deserted. Outwardly there is no sign of life. The colony is at rest.

Spring. As the sunlight warms again and the first flowers open the bees are encouraged to fly abroad more regularly. Their prime object at this time is the collection of pollen, needed for feeding their grubs, for the queen has again started to lay.

It is common in the early months of the year to see the workers moving from flower to flower, huge bundles of pollen "strapped" to each hind leg and often covered from head to toe in the stuff.

In January or February the queen gradually begins to increase her output of eggs in anticipation of the time when the nectar will be found flowing abundantly and the maximum number of bees will be needed. She is not alone in the task of rearing the young. In fact her only duty is egg-laying and once she has deposited the egg in its cell the young 'nurse' bees take over full responsibility of feeding and caring for the grubs during the first stages of their metamorphosis.

By mid-April the numbers are increasing rapidly and with the flowering of sycamore, willow, fruit blossom and dandelion, one of the major tasks of the year begins in earnest: the gathering of nectar.

As well as the rearing of workers, a number of drones will be produced in case a new queen should be needed and there is mating to be done. If the original queen is old and weak the colony will take steps to "supersede" her. That is to produce a young virgin queen who will mate and return to the hive to replace or help her ageing mother.

41

Another reason for queen rearing occurs should the colony become queenless. The bees are quick to sense the absence of their "monarch" and a state of great agitation ensues. They may remove several ordinary female larvae (which must be less than two days old) from their worker cells and place them in rapidly constructed queen cells. If all goes well a virgin queen will emerge, leave the hive in order to mate, and return, restoring order and harmony to the colony.

If the spring is kind, the queen prolific and nectar and pollen collection gets off to a good start, it will not be long before the colony, growing quickly, will be faced with the problem of insufficient comb. If there is space, additional comb will be built. If not, preparations will be made for swarming.

As in the previous instances queen cells will be built, eggs deposited therein with the appropriate nutrients, and the cells sealed. All is then ready for the dramatic exodus that provides a spectacle as enthralling as any event in nature. Leaving behind some young workers, a few older ones, the brood at different stages of development, and a number of maturing queen cells, the old queen quits the hive accompanied by a mass of workers. Sometime after this the first virgin queen hatches and quickly sets about destroying any remaining queen cells.

Swarming may also occur when the colony is attempting supersedure. In this case it will be one of the newly emerged queens who leaves with the swarm, the old queen remaining behind.

To an observer the first sign of a swarm is that a large number of bees appear around the entrance to the hive, behaving in an unusual and expectant way. What is going on inside at this time has been vividly described by Langstroth:

"On the day fixed for their departure, the queen is very restless, and instead of depositing her eggs in the cells, roams over the combs, and communicates her agitation to the whole colony. The emigrating bees usually fill themselves with honey, just before their departure; but in one instance saw them lay in their supplies more than two hours before they left. A short time before the swarm rises, a few bees may generally be seen sporting in the air with their heads turned always towards the hive; and they occasionally fly in and out, as though impatient for the important event to take place. At length a violent agitation commences in the hive;

the bees appear almost frantic, whirling around in circles continually enlarging, like those made by a stone thrown into still water, until, at last, the whole hive is in a state of the greatest ferment and the bees, rushing impetuously to the entrance, pour forth in one steady stream. Not a bee looks behind, but each pushes straight ahead, as though flying for "dear life", or urged on by some invisible power, in its headlong career." (Note 6).

As the swarm leaves the hive the sky becomes overcast, the air filled with the exciting vibration of many thousands of tiny wings, as the bees move back and forth at first uncertain of their goal. Soon the air will clear as more and more bees attach themselves to the cluster that will form on a branch, a post, or even on the ground, but never far from the hive though sometimes at a great height. There they remain in a shimmering mass in the midst of which is the queen.

Soon, some of the swarm will fly off in search of a suitable spot in which to make their new home, returning to communicate their discovery in the same manner that sources of nectar and pollen are communicated. Once a new home is occupied, comb is constructed, gaps filled and the structure generally made sound, and thus a new colony has been propagated. The reserves of honey consumed before leaving the old hive will keep the workers going until production is under way in the new premises.

Summer. With the passing of spring most of the major changes and upheavals in the hive should be over. Ahead stretch the summer months: a perpetual harvest of nectar and pollen that lasts (weather permitting) from dawn to dusk.

At this time of year the story of the bee is as much an inventory of the flowers she visits.

Borage, — one of the bees' favorites

Bees have a symbiotic relationship with flowers, established (we may conjecture) long before man was around to enjoy the fruits. It arose because bees and certain flowers adapted to each other. Species are able to do this because they are programmed genetically to be variable. Each generation of a species contains individuals more able to survive than others in a particular environment. The offspring of these survivors retain their parents' characteristics, with further variations which determine their own ability to survive. For this reason vastly different forms of life have evolved on our variable planet, from squat angler fish with lights dangling over their open jaws to soaring eagles stalking their prey on the wing.

Let's digress a little further. Botanists have determined that conifers evolved before any flowering plant. Types producing large quantities of pollen survived because the winds effected cross-fertilisation. Catkin-bearing plants evolved later, some of them with pollen attractive to insects, which equipped them with a secondary means of cross

44

pollination. As certain plants began producing less pollen they were able to use the resources thus saved to develop flowers which would attract even more insects. It is interesting that plants like magnolia, surprisingly one of the Buttercup family, are still visited by beetles for the sake of their pollen. Beetles are the most primitive form of insect, and Buttercups amongst the earliest flowering plants to have evolved.

Not all plants are pollinated by insects and wind. Chickweed for instance is mainly self fertilising, but generally speaking plants tend to thrive better on cross-pollination since this confers vigour on the offspring and improves their chance of survival. For example the plant known as 'Jacob's ladder' has stamens which ripen before the stigma is mature so that the stigma tends to be fertilised by pollen from other plants. In this way many flowering plants depend on their relationship with insects, as anyone who has beehives near an orchard wall will confirm.

In advertising their wares, pollen and nectar, some flowers have developed an array of features attractive to bees, the enjoyment of which we are able to share. For instance the first feature to attract the bee to a flower is its scent: a bee's ability to smell is said to be a hundred times greater than ours. Once the bee is within one metre colour becomes the most important influence. We can distinguish many colours; bees only four, for they cannot see far into the red end of the spectrum and are unable to distinguish red as a colour at all. On the other hand they perceive ultra-violet light waves as colour. Their favourites are yellow, purple and especially blue. The reason bees are attracted to white bryony and red poppy flowers is presumably because they reflect ultra-violet light.

As the bee moves still nearer, it is the form and patterns of the flower that draw her. It seems that bees prefer broken rather than regular outlines. Many flowers, such as violets and primroses, have markings which guide the bee straight to the nectary.

In order to guarantee cross-fertilisation a flower needs to provide the bee with as many grains of pollen as possible and some flowers have developed ingenious mechanisms for this purpose. Broom flowers have sprung petals: a bee alighting thereon cannot avoid a thorough dusting on its back and underside.

Harmonious Co-operation

Bee and Sage Flower.

Fig 1 shows the immature sage flower, (a) is the stigma (b) is the stamen with pollen boxes.

Fig 2 shows how a bee, in search of nectar, (located at the nectary shown at (c)) pushes the hinged stamen which brings the pollen boxes into contact with its back.

Fig 3 is a more mature flower, the stigma of which has elongated so as to receive pollen from younger flowers already dusted on to the backs of visiting bees.

The Results:—

(a) Cross pollination and therefore vigour conferred to the progeny of the plant, and

(b) Nectar for the bees (and honey for beekeepers!)

Most attractive to bees are small flowers in clusters, rather than solitary large blooms, and this makes sense in terms of the economy of the hive.

The interdependence of bees and clover is well known, white clover being totally infertile if no bees are present. Experiments in U.S.A. have shown that a clover field, where bees are kept, will yield twenty times as much seed as those without nearby colonies. Interesting, too, is the way in which the plant has adapted to the bees, by curving down any florets which contain no more nectar. The bees' efficiency is increased by not having to investigate empty florets. Efficient bees fertilise more flowers.

Nature's economy.
Clover florets droop after fertilisation, conserving the energy resources of plants and insects.

As the summer months pass, the bees move from the early fruit blossoms and dandelion to the hedgerows and trees — hawthorn and sycamore, holly and chestnut. In areas where soft fruit is cultivated, currants and raspberries are worked: brassicas, left to flower are favourites too.

With the arrival of mid-summer, the clover is in full bloom; lime trees, too. Blackberry and willowherb keep the bees busy until the final crops of heather and ivy wind up the honey season. Apart from these major honey plants there are inummerable other species that, although usually found in less profusion, are no less attractive to bees.

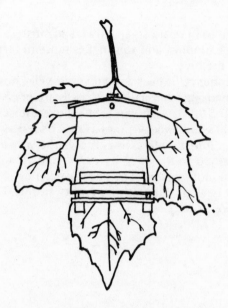

Autumn. By the time the heather and ivy flowers have
finished the temperature is usually quite low and those
milder days that bring the bees forth from the hive are again
as rare as they were in February and March. From now on
the colony must rely entirely on the stores it has been putting
by during the summer. The queen, whose egg-production
was at its height in the spring so as to build up the colony
for nectar collection, has by this time almost ceased laying;
and consequently the number of workers will gradually have
decreased. With all the frantic activity of the summer
the worker's life expectancy is about five or six weeks, but
those who are reared towards the end of the summer will
live right through the winter, the only energy required being
to keep warm. It is these bees who will rear the first brood
of the spring.

As well as the reduction of the worker population that
occurs naturally in preparation for the winter, the workers
themselves take steps at this time to dispose of any remaining
layabout drones, whose usefulness has long been outlived.
They are first deprived of nourishment and then, when they
are thoroughly weak, they are dragged by the workers from
the hive and cast upon the earth to die.

The colony is now ready for winter.

PART TWO:

Beekeeping

The Apiary

he first consideration, once you have decided to keep bees, is *where* to keep them. Your decision will be influenced by how many colonies you intend keeping. Remember, too, that if you are successful and enjoy the activity you will soon find your apiary expanding. When choosing a site for your hive, there are certain factors that *must* be taken into account and other factors which should be considered if the best results are to be achieved.

Most important is that the bees should have sufficient space in front of the hive to allow them freedom of flight. Don't place them in the middle of a thick and overhanging wood. Don't let the grass grow tall in front of their entrance. A board placed on the ground will keep the immediate vicinity clear. Don't put them in such a position that they have to bombard your neighbour's garden path on every flight. If you are using a fairly well populated area, or if there are farm animals about, a hedge or a fence around the apiary is a good idea. It both protects the hive from interference and also, by forcing the bees to elevate their flight path, it protects passers-by from the danger of being stung. In law you are responsible for your bees' actions!

In general an airy, partly sunny, partly shady location on firm ground is desirable, though in England and similar latitudes shade is rarely necessary! Your bees will be encouraged to leave the hive only when it is warm enough, so don't discourage them by placing the hive in a very windy spot, hidden from the morning sun. A sheltered position on a south facing slope is ideal — the entrance of the hive should face towards the sun and away from the cold north winds. It has also been suggested that hives should slope downwards slightly towards the front. This is to help drain any excess moisture and to assist the bees in the removal of dead bodies and other debris.

Remember, in a warm hive bees start early and finish late, especially in spring and autumn which are critical times for the success fo the colony.

The worst hinderance is damp. It will quickly rot your equipment. More catastrophic is the effect of damp upon the bees. Firstly they must struggle day and night to maintain a suitable atmosphere within the hive, thus they are continually wasting energy and accumulating stress. Such a hive will be short of honey, short of temper, and sooner or later will become diseased. For these reasons, don't put your hive in that vacant patch of the garden that isn't fit for anything else. If nothing will grow there, certainly your bees aren't going to flourish.

If you are planning to keep just one or two hives then you don't have to worry much about sources of nectar and pollen. As was mentioned in the introduction bees have been successfully kept in the middle of London. Bees will happily forage to two miles and if pushed will travel ten. Most areas have sufficient nectar-producing vegetation within that range. However as your apiary grows in size the availability of nectar and pollen becomes a real consideration and of course if the bees must fly only one hundred yards on each trip much energy and time is saved. A rough guide: an acre of well stocked ground or at least four acres of sparsely stocked ground can support a colony.

So, don't get carried away too soon. Unless your area is very rich in nectar-bearing flora, a couple of hives properly sited and well managed will provide far greater return in surplus honey than a dozen carelessly managed and neglected colonies.

Another important feature is a continuous supply of water nearby, with a suitable perch from which the bees can drink (a partly submerged stone). Ensure that the water is fresh; stagnant water may contribute to the transmission of bee diseases.

That takes care of most of the bees' needs. There are yet a few details which make life easier for the beekeeper. Leave

room by the hive in which to work: about one square metre
is sufficient. You must have somewhere to put the "lifts"
and "supers" as you remove them. Hopefully the time will
come when you are going to harvest a sizeable honey crop.
A "super" full of honey is not going to weigh much less
than 20 lbs and may weigh a lot more, so you won't want to
carry it very far. For this reason, and because beekeeping
always involves you in handling a fair amount of equipment,
keep your bees near your home, or have a shed nearby in
which to store all the gear, or at least place the hives in such
a spot that provides easy access for transport. One further
advantage of having your apiary near to home is that you are
more likely to be around when you are needed, and there
is less chance of the bees being forgotten and neglected (out
of sight, out of mind).

Bearing in mind all these basic requirements you will soon
notice suitable spots for positioning your hives. Whether it's
a flat roof, a vacant city lot, an orchard, or a backyard,
almost certainly there's a corner just waiting for a beehive.

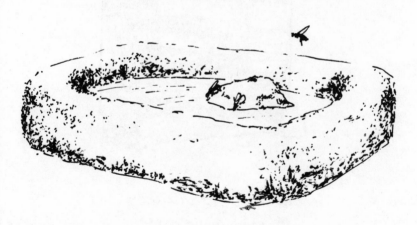

almost certainly
there's a corner
just waiting for
a bee-hive.

The Means

The Beekeeper

here are great differences of opinion between the experts on the question of the bees' sensitivity to the state of mind of the keeper. However, once you begin handling the creatures for yourself you will soon discover that they do not behave in the same way each time you visit them. Although there are a number of different factors that influence the mood of the colony, the mood of the beekeeper cannot be ignored.

As is the case with most creatures, bees will be made angry by treatment that is rough and unfeeling. Certainly if the bee keeper goes into the fray hidden in a thick and pungent smoke screen, he can be clumsy and brutal and yet emerge unscathed. But this is not our approach. Our bees are going to be disturbed as little as possible; we are going to be gentle to them, sensitive to their changing moods and needs, and, softened by our love, they are going to become our friends. Yes? But let's not get carried away too quickly: however expansive our feelings towards the bees we must begin with a rudimentary knowledge of the other things that effect the temperament of the hive, otherwise we may come away disappointed and hurt, both emotionally *and* physically, by our first encounter.

On a day when the weather is foul — cold, windy or wet — not only will all the bees be at home, but they will be slightly irritable: they cannot go out to forage. If, at such a time, the roof is removed from their home, letting in cold air and bringing with it the scent of danger, one should not be surprised if they react in an aggressive manner. From this we can deduce that the best time for working inside the hive will be in the middle of the day, when the air is warm, the sun shining and the wind gentle. At such a moment the bees busy, contented and many are abroad.

There are other conditions that can produce a bad-tempered colony, even on sunny days. (Let us forget for the moment the possibility of having a naturally bad-tempered strain to start with.) Examples are a heavy atmosphere, prior to thunder; of the cessation of a particular honey flow, which can occur as a result of a fall in air temperature: the bees get upset by this because they continue to fly to the spot they have been working all day to find that all of a sudden there is no nectar available.

Very dry conditions, if they are prolonged, can also put a stop to the honey flow. It should be sufficient to say that any condition that imposes a stress on the colony will give rise to bad temper and likelihood of attack if disturbed.

Having become familiar with all these outside influences that can be at work on the stability of the colony, one is then in a better position to judge to what extent one's own feelings and behaviour will affect the bees. Certainly they can sense fear and, just as with a dog, the more they feel your fear the less they like you. For this reason in the early days, if you are a bit unsure of yourself, it is best to be *completely protected:* wear a veil and gloves and an overall tucked into boots. This is a great help in building up confidence, then when you feel more at home and you have become used to the buzzing in your ears, you can strip off a bit.

The way to inspire your bees' confidence is firstly to be sensitive to what's going on in the colony at any particular moment, and then to act with firm and gentle movements, knowing just what it is you wish to accomplish and doing so in the shortest possible time, but not hurrying and never panicking.

If your idea is to work with your bees, rather than to have them work for you, then the more time you can spend with them the more likely they are to become familiar with your particular vibration, and the sooner they will learn that they have no need to get into a state when you are around.

Rushing out to the apiary once a month, breaking into

their home to see how much honey they have made for you
to take, is not the way to establish mutual trust.

Equipment

"The first decision the embryo beekeeper must make is the
kind of hive to use. In his selection he is sometimes misled by
cranks, of which unfortunately, there are a few whose pride
and boast is that they are against orthodoxy. A mistake at
the commencement will ultimately cause annoyance,
together with much expense and trouble, difficult to rectify."
W. Herrod Hempsall.

The earliest hives were no different from the homes of
wild bees: often hollow logs with a simple arrangement of
crossed sticks at the top to support the combs. Later straw
skeps were developed and various systems of boxes upon
which extra boxes could be stacked to supply space for
surplus honey. But with all these methods there was no way
of harvesting the honey without driving the bees away or
killing them: not only unfair but unprofitable.

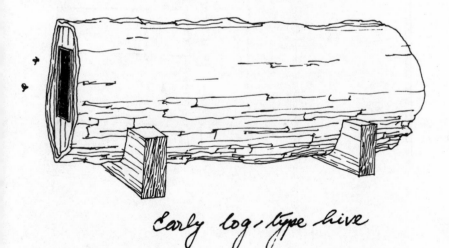

Early log-type hive

On October 5th, 1851, the Reverend Lorenzo Lorraine Langstroth patented his invention of a hive that opened at the top and which contained ten moveable frames in which combs were suspended and which could be removed without crushing the bees. The basis of the invention was Langstroth's discovery of the "bee space" — a gap of about eight millimetres between all the surfaces. This occurs naturally in all colonies and once discovered, like most things, is rather obvious, being the space required for easy passage of bees around the hive. Since those days various different designs (using the same principle) have been developed. When deciding on the best kind of hive to use you will be confronted with a number of different choices, the two most common of which are the single walled or National hive and the double walled or W.B.C. hive. Each design has

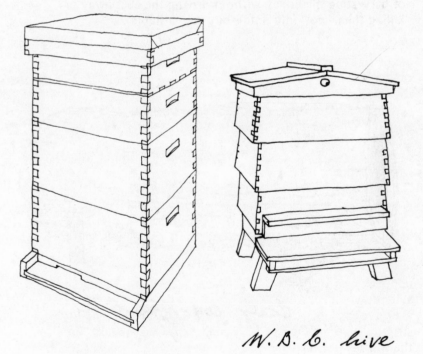

W. B. C. hive

National hive

its opponents and those who will happily lay down their lives in its defence.

Those who use single walled hives will tell you that they are cheap to construct, simple to handle, suitable for all the bees' needs and that their only disadvantage is that they don't have that English country garden appearance of the W.B.C. type.

Let us turn again to the opinions of Mr. Herrod Hempsall to learn of the advantages of the double walled hive.

"With over fifty years' experience in beekeeping in all parts of the British Isles, we assert most emphatically that the best kind of hive to use is the double walled one known as the 'W.B.C.' hive, the cognomen being derived from the initials of the designer, the late William Broughton Carr. The advantages possessed by this hive over any other are its simplicity and mobility; its accessibility for cleansing and disinfection; the cavity between the living quarters and the outer wall which prevents sudden fluctuation in temperature within, in response to sudden and violent exterior climatic changes, and, most important of all, the circulation of air in this cavity keeps the home absolutely dry" (note 7). There are other systems. Whichever design of hive you decide upon the basic principle will always be the same: at the bottom is the *stand*, or *floorboard*. This is a horizontal platform upon which the hive stands, and it should be raised clear from the ground to allow for passage of air and freedom from damp and intruders.

Placed on the stand is the *brood chamber* at the front of which is an entrance. This is the box where the queen lives and lays eggs, where workers store the food necessary to nourish the larvae and where the young bees emerge from their pupae. Opinion varies as to the optimum depth for the brood chamber, try one deep box and one shallow together.

Above this is placed the *queen excluder* which, as the name suggests, is a device for excluding the queen from the upper part of the hive, thus ensuring that her egg-laying activities are restricted to the brood chamber. There are two main types of excluder on the market, one being made from strips of wire held within a wooden frame that fits on top of the walls of the brood chamber. The other is a simple sheet of perforated zinc. In both cases the gaps are large enough to allow the workers through but not the queen or the drones.

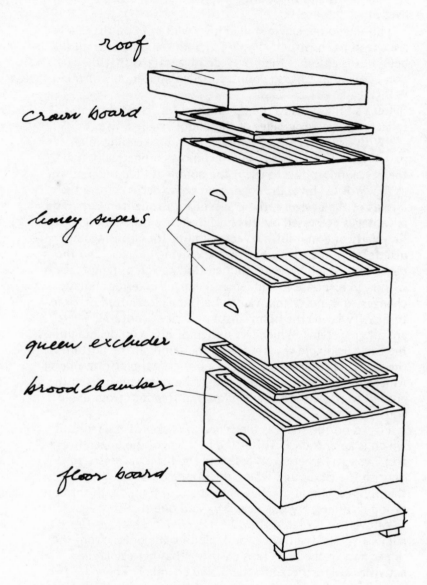

roof

crown board

honey supers

queen excluder

brood chamber

floor board

60

It is best to choose a queen excluder which provides bee space between it and the top of the frames on which it is placed.

Section of queen excluder

Above the queen excluder are placed the *honey supers,* boxes containing combs just as in the brood chamber, but it is here that the worker bees store their surplus honey. Any number of these, within reason, can be placed one on top of the other as space is required.

Above these is placed some kind of *cover* forming the inner extremity of the bees' home and keeping out the cold. If for a cover you use a wooden board then this can also be used with feeders, and as a clearing board (see below) provided it has the appropriate aperture in the middle. Sometimes glass or perspex is used which is nice for looking through but invites condensation, whereas wood allows the hive to breathe.

Lastly, on top of this is the *roof,* watertight and firmly positioned to keep out the elements.

This basic equipment provides the bees with all they need for a comfortable life, with the brood area separated from the honey store.

In order to "harvest" the honey some means of encouraging the bees to vacate the top honey-filled supers is needed. For this a *clearing board,* fitted with a *"Porter Bee Escape"* is the easiest and kindest way. The supers which are to be removed are lifted off the hive for a moment, the clearing board placed on top of the remaining boxes and the supers replaced above the board. The bees then move down to the lower part of the hive and cannot return. Within 24 hours the supers are usually empty of bees. A more brutal method is to lay a cloth saturated with some chemical (in the past carbolic was used, today concoctions with much longer and unpronounceable names are recommended) on top of the honey super driving the bees rapidly down and away from the cloth. Smoke can also be used but is not very effective.

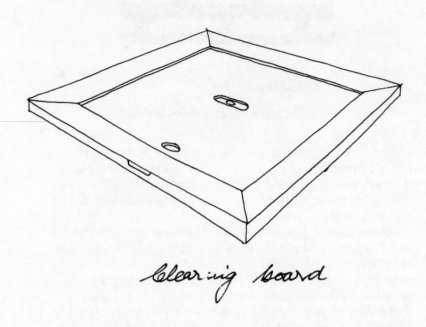

Clearing board

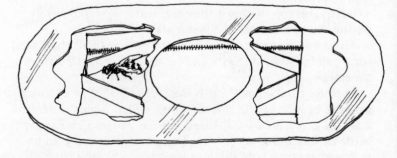

Porter Bee, escape cut away to show brass springs.

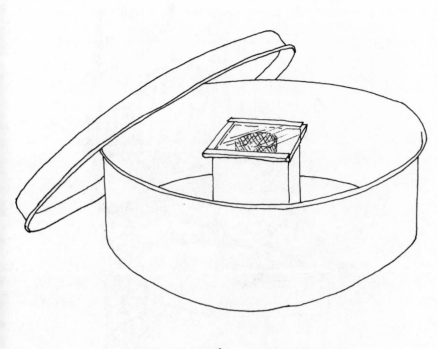

Feeder

This same clearing board, once the bee escape has been removed, can be used, at various times when the colony has to be fed, usually in the autumn or spring. If the board is placed on the top super, a feeder can be rested on it and the bees are able to pass upwards into a feeder while the hive remains insulated from the cold. The best type of feeder on the market is pictured here, but a perfectly acceptable type can be made from a seven pound honey container (or similar receptacle) by piercing a dozen or so holes in the lid, turning it upside down and placing it over the opening in the board (with the bee escape removed).

The beekeeper is advised, at least to begin with, to wear some form of protective clothing. A *veil* is essential to give one confidence, *gloves* or gauntlets are a help too, though experienced beekeepers tend to discard them as they make

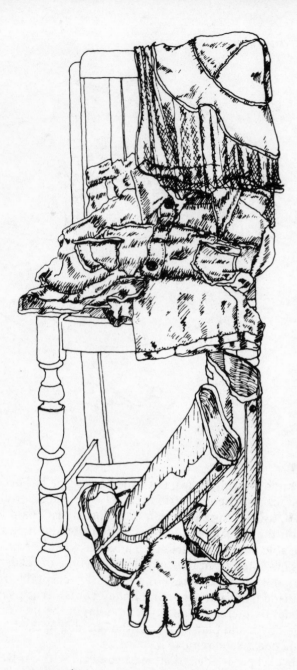

When you've got used to the buzzing
in your ears, – you can strip off a bit.

for clumsy handling. An *overall* tucked into *strong boots* completes the defences.

The only other essential items for field work are the *hive tool* and the *smoker*. The hive tool is used for easing the moveable components of the hive apart, very necessary as the bees do an excellent job of sealing all the gaps and crevices with propolis. This implement can also serve for scraping and cleaning parts of the hive that may become obstructed with comb and propolis. The smoker is used to quieten the colony if it begins to turn nasty. Smoke has the effect of making the bees react as if the hive were in danger of fire; they begin to behave as if they were soon to leave their home, that is to say they fill up with honey in preparation for the trip. This has a soothing influence on them and like all creatures they become less aggressive

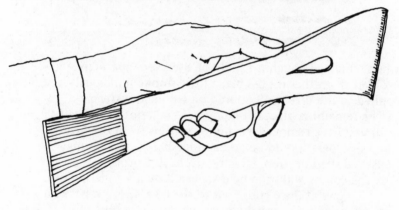

Hive tool

Smoker

a little smoke, judiciously applied, is oil poured on troubled waters.

when replete. Though it should be used very sparingly and avoided altogether if possible, a little smoke judiciously applied to the hive is oil poured on troubled waters.

The remaining pieces of equipment will not be needed until you have removed the honey surplus from the hive. The full supers should be taken immediately to a "honey house", a shed or room which must be bee-proof, otherwise the combs will soon be discovered by a wandering bee and the news at once conveyed to the hive. It is a mistake you will only make once! Unless you intend to cut the combs bodily from the frames you will need an *extractor* to remove the honey from the comb. There are various designs, some made of tin or stainless steel, and others plastic; some hand-operated and some with motors, but the principle is always the same. The combs are spun at speed, forcing the golden liquid to fly out against the sides of the drum, down which it trickles, settling into the sump where it can be drained by means of a tap. An extractor, even a manually operated one, is an expensive piece of equipment for the small scale backyard beekeeper. Try to locate other backyard beekeepers and share with them the use and cost of such a machine. Alternatively your local supplier may offer an extracting service at reasonable cost, or even hire you an extractor.

66

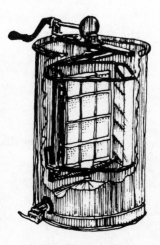

extractor

A useful device is the *solar wax extractor*. Let the sun provide all the energy needed to render any old pieces of comb down into blocks of clear golden beeswax. Such blocks can be traded against new equipment from the suppliers or kept for wood polish or home manufacture of foundation.

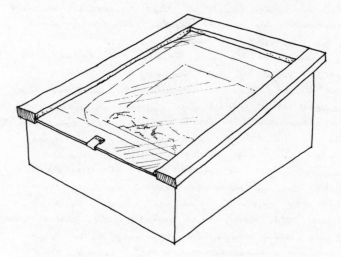

Solar wax extractor

Making a Solar Wax Extractor.

(A) is the box, made of marine ply or other
weather resistant sheeting, strengthened
at the corners by (B) battens app 25mm square.

(C) is the glazing framework, the bottom rail (D)
being cut down to allow the glass to overhang.
There are two sheets of glass, supported on and
separated by small section beading (E) and
puttied at the top edges.

(F) is a lead tingle which helps to secure the
bottom edge of the glazing.

(G) and (H) are handy containers (biscuit tins,
baking tins, etc.) you happen to have, so
no dimensions have been given, - you build
the extractor to suit the size of your
tins, punching holes in tin (H) and
supporting it with more timber battens (I).
Both tins should be loose, for easy accessibility,
and the cover hinged and close - fitting.
Many other designs to achieve the same result
have been perfected. You may like to do
some more research yourself, using this
one as a guide.

Making Equipment

Having mentioned the equipment one needs to get started, we should examine the cost of it all and how it can be obtained.

The first alternative is to buy new from a beekeepers' supplier. This is easy, reassuring and expensive. A few suppliers who will despatch their wares nation-wide are listed at the end of the book, but there is sure to be a small local business near you who, being agents for the manufacturers, can give as good a deal as you will get by sending off direct.

It is possible to obtain all hive parts either ready made-up or "in the flat": that is to say all cut to size but not nailed together: obviously this works out a little cheaper.

If you don't fancy the idea of buying new equipment you can either try to buy secondhand from a retiring beekeeper (a rare species!) or set about making your own. Certain items such as the hive tool and smoker would be difficult to make though an old chisel and a piece of smouldering sack carefully handled will serve well enough.

Finding secondhand hives can be a most enjoyable and rewarding enterprise! One occasionally sees what appear to be clusters of neglected hives in an old overgrown orchard or back garden. A polite enquiry can at worst leave one assured that the hives are in fact being cared for, but more often it will lead to a fruitful discussion on beekeeping and other country matters and produce a hive of bees at little or no cost.

There are quite a few instances where the beekeeping enthusiast in a family passes on leaving his relatives with an unwanted responsibility. An advert in the local paper may jog their memory; also your local supplier may know of some secondhand gear. It is difficult to put a price on such equipment: the author's first hive was acquired from a kindly old gentleman in Somerset for nothing more than a smile and a promise that they would be cherished. Two well-established hives were sold recently at a local auction for little less than the equivalent hives new.

Now we come to the question of making your own. There is certainly a saving made this way, though if new timber is used the hives will still prove far from cheap. Old wood may safely be used for the outer parts of the hive, but care should be taken with the supers and brood chamber and the timber thoroughly cleaned first.

The following diagrams will help, two points being worth remembering: due to the absolute necessity of maintaining the bee space throughout the hive, these measurements should be followed closely. Also always stick to one design of hive so that all the pieces in your apiary are interchangeable.

The Double Walled Hive (W.B.C. Type)

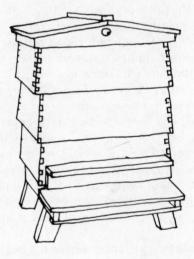

W. B. C. hive

When building this type of hive the outer parts (lifts) don't have to be splayed which, while looking very pretty and "traditional", makes for rather complicated carpentry. They can be box shaped with a lip around the bottom edge to fit neatly onto the top of the lower lift. (Note 8).

It is doubtful whether there is much advantage in building your own frames for comb. Firstly the measurements need to be so precise and the woodwork so finicky that only an experienced joiner would have much success. Secondly, readymade frames can be bought 'in the flat' very reasonably and nailed together in a few minutes. Once you have your beekeeping venture off the ground, and if you fancy a bit of intricate woodwork, many winter hours can be happily spent constructing your own deep and shallow frames. Just in case; here are the plans.

You will continually be needing new sheets of foundation for extra supers and for replacing old comb. Every year the price of foundation rises enormously and this is one piece of equipment that you might think you would have to buy. But here is a straightforward method of making your own.

You will need: 3 metres of 25 mm by 25 mm timber
a small quantity of portland cement
twenty 75 mm nails
two butt hinges
one sheet of machine-made foundation
also you need to save any old wax from cappings or broken comb and this must be rendered down as described on page 72.

How to make the mould *(see illustration page 81)*

1. Make two equal-sized rectangular frames of 25 mm x 25 mm wood. The inside measurement to be the size of the foundation required.
2. Place the frames squarely one over the other and fix with two butt hinges on one of the longer sides, so that the frames open and close like a book.
3. Drive 75 mm nails into the inside of the frames to act as keys for the cement.
4. Open the frames out and place face downwards on a level surface. Fix a sheet of machine-made foundation firmly under one side.
5. Brush thin cooking oil over the foundation to prevent the cement from sticking.
6. Pour on cement — mixed to a cream without sand and sieved to avoid any lumps. Fill the frame carefully to avoid any air bubbles and leave to set.

7. When set, close the frames together with the foundation still in position and place with empty frame upwards. Brush with oil.
8. Pour on cement and leave to set.
9. Open frames and remove foundation. Tidy up . . . this is your mould.

How to make foundation

1. Stand mould in a tray of shallow water in order to catch any over-flow of wax. The mould should be warm and this can be achieved by pouring hot water over it or placing it in a warm oven.
2. Coat mould with soapy water, removing any surplus.
3. Pour melted wax *quickly* on to one side of the mould and close mould for about one minute.
4. Open and remove sheet of foundation, trim. Sponge again with soapy water for each sheet.

These sheets can be fixed into frames in the normal way. Obviously your first attempts are going to be far from perfect, but so long as most of the area of the frame is filled with foundation the bees will do the rest and an even spacing of combs will be maintained.

Foundation made in this way will not have any wire supports, as does most bought foundation these days. This can be a drawback if you are going to extract your honey on a high speed machine, for the combs may collapse under the centrifugal force. Also very high temperatures in the hive may cause combs to sag, if they are not supported. However, unwired foundation can be an advantage if you plan on simply cutting up the comb and eating it, wax and all.

A glance at the current price of a sheet of wired, machine-made foundation should be enough to spur you on to try and make your own.

A solar wax extractor can be made easily enough. A box, similar to a cold frame, is fitted with *two* sheets of glass (in the usual manner for double glazing) and placed in direct sunlight facing south. Inside this frame is placed a shallow tray, sloping slightly, and perforated along the bottom edge. Beneath these perforations is a dish. The old comb is placed in the sloping tray. The melting wax runs down the tray and drips through the holes into the dish — any foreign matter remaining behind.

roof

lift

batten

supers

top of
floorboard

W.B.C. alternative

roof is felted, lifts square-
sided and battens nailed
to lifts to ensure a fit
and give the beekeeper
a grip for lifting

HOW TO MAKE YOUR OWN
W.B.C. TYPE BEEHIVE.

PLEASE NOTE :-   1. Drawings are <u>not</u> to scale

   2. All dimensions are in millimetres

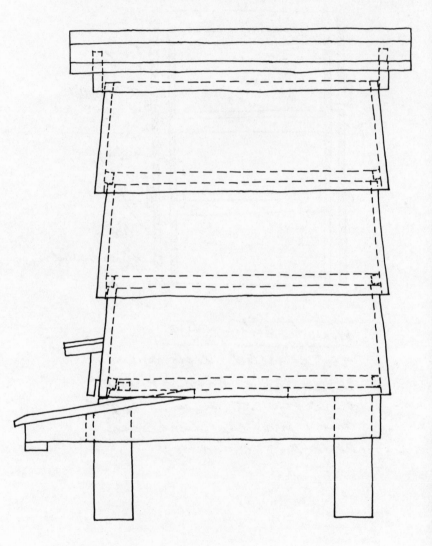

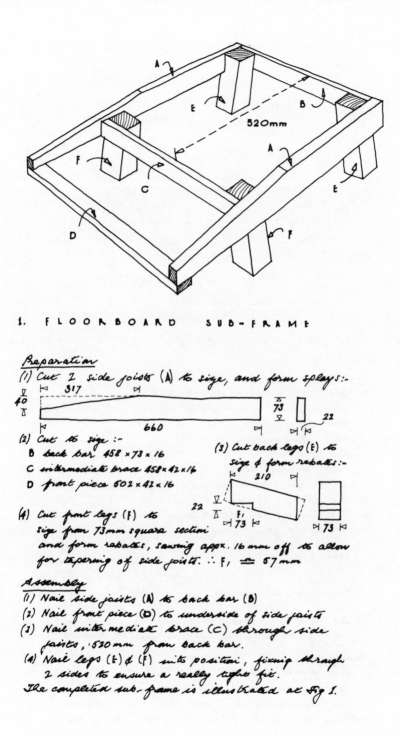

1. FLOORBOARD SUB-FRAME

<u>Preparation</u>

(1) Cut 2 side joists (A) to size, and form splays :-

```
  40    |←  317  →|                              73      |
                                                 ↕     ← 22
        |←──────── 660 ────────→|
```

(2) Cut to size :-

  B back bar 458 × 73 × 16
  C intermediate brace 458 × 42 × 16
  D front piece 502 × 42 × 16

(3) Cut back legs (E) to
size & form rebates :-

```
        |←  210  →|
   22                  F₁
        |← 73 →|          |← 73 →|
```

(4) Cut front legs (F) to
size from 73mm square section
and form rebates, sawing appx. 16 mm off to allow
for tapering of side joists. ∴ F₁ ≃ 57 mm

<u>Assembly</u>

(1) Nail side joists (A) to back bar (B)
(2) Nail front piece (D) to underside of side joists
(3) Nail intermediate brace (C) through side
    joists, 520 mm from back bar.
(4) Nail legs (E) & (F) into position, fixing through
    2 sides to ensure a really tight fit.

The completed sub-frame is illustrated at Fig 1.

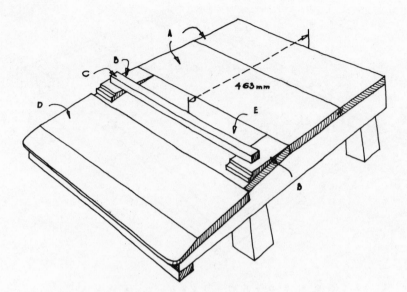

2. F L O O R B O A R D

Preparation

(1) Cut 2 floorboards and loose tongue (A) to size,
each 503 × 170 × 16, to butt at edges of side joists, back
bar, & point where side joists taper.

(2) Cut 2 tapered wedges (B)
and form rebates. NOTE
that 'vertical' cut of
rebate is splayed slightly (about 3 mm)

158   13

176   73   13

(3) Cut slip (C) to size 468 × 22 × 16

(4) Cut landing board (D) to size, form splay and
round off front corners

16

540   16   16

140   10   10   200

(5) Cut inclined board (E) 503 × 200 × 16
and form splay to suit horizontal boards

Assembly

(1) Nail floorboards (A) & (E) to sub-frame

(2) Nail tapered wedges (B) to inclined board (E)

(3) Nail landing board (D) and slip (C) centrally in
position, slip to be 463mm from back edge of board.

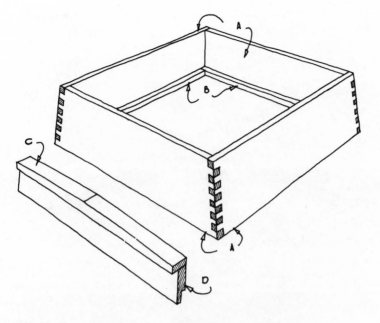

3. OUTER BODY

<u>Preparation</u>

(1) Cut 4 sides (A) to size and form splays :-

503

198

540

13

This could be the most awkward bit of joinery, so if you don't feel up to it, try one of the proprietory methods of end-jointing.

(2) Cut 4 spacing fillets (D) to size 495 × 22 × 16

(3) Cut porch roof (C) to size and splay back edge & top twice

22   6   250

73   500   16

(4) Cut porch (D) to size, splay top & rebate back edge

10   73   22   73   25   500

<u>Assembly</u>

(1) Nail sides (A) together
(2) Nail on spacing fillets (B)
(3) Nail on porch to line flush with underside of body, and nail on porch roof (D). For (2) & (3) see detail

Make 3 outer bodies, one with porch

C   A
D   B
16mm

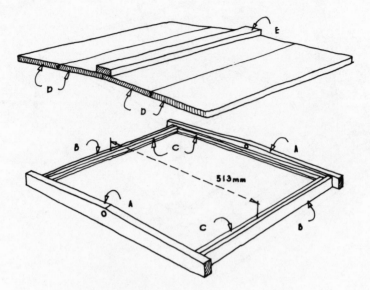

4. ROOF

<u>Preparation</u>

(1) Cut 2 end gables (A) to size and form splays :-

575

38

22

73

35

(2) Cut 2 side pieces (B) to size 513 × 35 × 16

(3) Cut 4 spacing fillets (C) to size 500 × 19 × 13

(4) Cut roof boards (D) and splay at ridge, — round
off outside corners. These can be 2 no. ply sheets
each 630 × 300 or lapped boards or butt-jointed boards
to cover the same area. Ideally if you're not going to
cover the roof with felt, the boards should be cut
from 4 t & g boards, each 630 × 150 × 16

(5) Cut ridge piece and form splayed rebate to
accommodate angle of roof boards :-

630

25

50

<u>Assembly</u> (NOTE : Drill 2 no. 25mm holes in gables for brass cones)

(1) Nail side pieces (B) through end gables 513mm apart

(2) Nail spacing fillets (C) to line in
with top edge of side pieces (B)

C

B

(3) Nail on roof boards (D) and ridge piece (E).

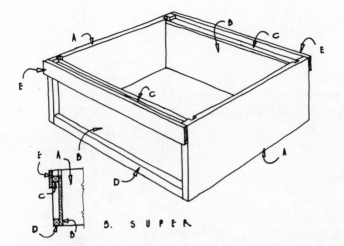

5. SUPER

Preparation

(1) Cut 2 rebated side pieces (A) to size, and form housing:-

$$4 \quad \boxed{} \quad 16$$
$$19 \quad 368 \quad 13 \quad 225(A_1)$$

NOTE: Measurement A, should be increased to 150 for shallow supers

(2) Cut plain side pieces (B) to size 398 × 208 (or 133) × 13

(3) Cut top & bottom packing pieces (C) & (D) to size:-
390 × 19 × 16

(4) Cut grips (E) to size 422 × 44 × 11

Assembly

(1) Fit plain side pieces (B) into rebated side pieces (A) and nail together

(2) Nail top and bottom packing pieces into position

(3) Nail on grips

Try making a couple of each type to start with

Now your hive joinery is complete, remember you have to provide ventilation for which 2 brass cones per hive can be bought from suppliers. You will probably also want to buy metal spacers to nail to the top edge of each plain side piece. This ensures a bee space is provided when the comb foundation is in position. Apart from this, you will have made your hive using only your own nails, tools, wood,— and patience!

79

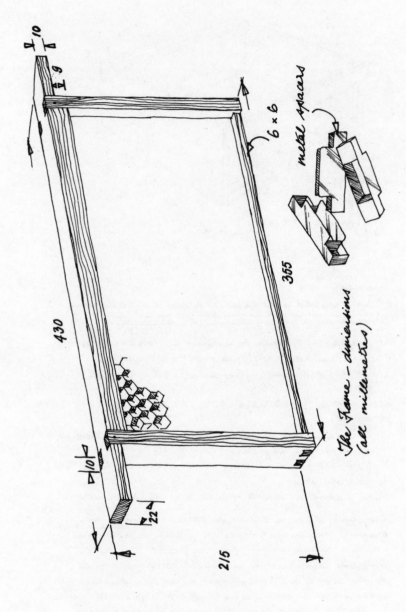

The Frame — dimensions
(all millimetres)

metal spacers

6 × 6

430

365

215

22

10

10

9

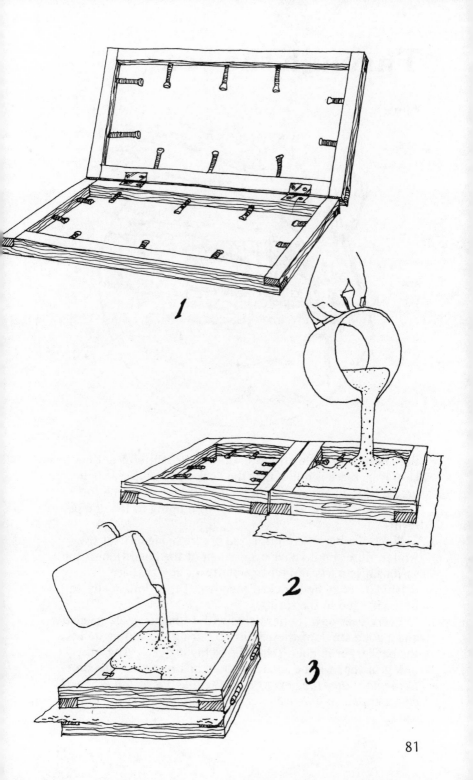

Through the year

Winter

Bees need very little attention during the winter months. However, there is much that the beekeeper can achieve in preparation for the following season. This is the time for taking care of any work that may need doing to the equipment.

When the bees were inspected before being left for the winter, note should have been taken of the condition of the various hive parts and replacements for any that are delapidated can be knocked together at this time, ready to be exchanged in the spring.

Every year new frames fitted with foundation are needed, and a good stock of these built up during the winter stands the beekeeper in good stead the following year. Any other new gear that will be needed should be ordered at this time. *Remember that the new price lists come out in January and an order placed in November will secure goods at the old price.*

The occasional visit to the apiary will be reassuring. Not that any bees are likely to be seen, but there is always the possibility of a damaged hive due to stormy weather, other creatures, or even human vandals. If the bees were properly prepared for winter the beekeeper need not much concern himself with the occupants at this time, unless inclement weather continues long into spring. Then there is the likelihood that supplies will begin to run low.

During mild, sunny weather, cautious inspection inside the top cover may be made in the middle of the day. If there is little capped honey visible, or the bees are clustered close to the top of the hive, they must be quickly fed. If the weather is continually cold, the hive should never be opened but if concern is felt, an estimate of the amount of honey left can be made by carefully lifting the hive. If it moves with little effort then immediate action is necessary.

Opinion differs as to the quantity of honey that should be left with the bees for the winter. Many beekeepers will leave one completely full super (about 30lbs) and expect to feed the bees again in the spring. To avoid this and to be sure of maintaining a strong and healthy colony right up to the start of the nectar flow, it is best to see that colonies kept in the British Isles or similar latitudes have between 50lbs-60lbs at the onset of winter. Often the bees will not need this much in which case nothing will have been lost.

This store will be gradually consumed and by the time the nectar is flowing steadily again, perhaps only 10lbs-20lbs will be left. If at any time you reckon there is less, which may happen if the colony is large and the weather unkind, the bees must be fed.

The most common method is to place a feeding tin over an opening in the centre of the inner top cover and provide syrup (see page 63). Many people recoil from the idea of taking the bees' honey and replacing it with white sugar and ideally enough of the bees' own honey should be left for their needs. However, in the event of an emergency, substitutes made from white sugar can avert a disaster. Don't be misled into using brown sugar (which will give the bees diarrhoea) or honey from some other source (which is an effective way of passing on disease to your hive).

As early as February, the queen may be laying again. For brood rearing two provisions are necessary: pollen and water. Pollen, the bees will take care of by themselves. Water however, should be provided nearby in the form of a bowl or dish, also some means for them to land and drink without drowning. A partially submerged stone will do. Make sure the water is pure and fresh.

In early spring, some commercial beekeepers will provide extra syrup and artificial pollen (made with soya bean flour) to stimulate brood rearing.

During the spring the changing environment outside the hive must be carefully observed. If the weather is mild and blossom profuse the colony will need no feeding, but continuing wet or cold weather in spring and scarcity of blossom can combine to threaten the life of the hive. If little food is coming in while the population is rapidly increasing the situation can change from secure to hopeless in just a week or two. In such instances regular feeding is necessary.

About mid-April very weak colonies may be united. It is much more profitable to have one strong colony than two weak ones for it is the surplus we are after! Uniting is best carried out as follows: all parts of the hive are removed except for the brood chamber containing the bees and the stand upon which it rests. A sheet of newspaper is then placed directly on top of the frames. In early morning or late evening, when the bees are not flying, the brood chamber of the second hive is fitted and placed on top of the first. The two brood areas are thus separated by the sheet of newspaper. Care must be taken that there is no way the bees in the top box can leave except by tearing a passage through the paper. This they will do within a few hours, by which time the two colonies will have become accustomed to each other's presence and continue life as one. The two queens, however, will not be willing to share their matriarchial role and will have a fight about it (to the death), the stronger and fitter winning. The observer can tell when the union has been effected, because small particles of paper appear at the entrance and bees can be seen leaving the hive dragging scraps of last week's news behind them.

Once the weather is mild and settled, a warm day should be chosen for a thorough inspection of the hive. First a little smoke is applied at the entrance and after a moment or two the top cover removed. All the brood frames should then be scrutinised one by one, and any that are broken or decaying should be replaced. Presence of queen cells should be noted and, depending on your approach to the matter of swarming, removed or left where they are. The stand must be checked for level; the outer parts checked and replaced where necessary; any cobwebs or debris brushed away and the whole reassembled and left with a queen excluder on top of the brood chamber and a super above this. All these operations and any observations on the condition of the hive should be recorded in a notebook for future reference.

Since it is in the beekeeper's interest to keep the drone population to a minimum, this opportunity should be taken to remove any combs containing drone cells that may be present in the brood chamber.

Thus the beekeeper ensures that as the time of the main nectar flow draws near, his colonies are strong, and as well provided for in terms of food and shelter as possible.

Any time from the end of April until July swarming may occur. This can be the most exciting and fascinating aspect of keeping bees and also the most tiresome. We'll first describe the methods of capturing and utilising a swarm should it occur. Then we'll look at various ways in which swarming can be prevented.

Such a commotion arises when a hive begins to swarm that if you are anywhere within a couple of hundred yards you will almost certainly be aware of it. As the sky blackens and the air is filled with the bees' collective hum, there is nothing one can do but stand and enjoy the spectacle. There is really no need to wear hat and veil or other protection as to be stung by swarming bees is very rare (though not unheard of!).

Hopefully someone will have noticed the proceedings at this stage for, within a few minutes, the swarm will start to settle in one place, on a branch or post or the side of a building, and before long all will be still again and the swarm easily missed.

When all the bees are collected in one place, you can go into action. You will need a box, open at one end, and possibly a small, soft carpet brush.

Ideally the swarm will be hanging from a low, easily accessible branch. In this case you can place the box beneath them and give the branch a sharp shake, whereupon the bees will fall — as one — into the box. This should then be placed upside down, quickly, upon level ground, and propped open at one edge to allow entrance and exit. If the day is very hot it is best to shade the box.

Unfortunately your first swarm will probably settle some-where quite inaccessible or close to the ground on the trunk

*ideally the swarm
will be hanging
from a low, easily
accessible branch*

of a large tree. In such an instance, if there is room to place a
box beneath the greater mass of the swarm, success can still
be achieved by brushing the bees into the box with one or
two swift, sure strokes. Again the box must at once be placed
face down on a board or level ground. If, however, there is no
way of positioning a box underneath the bees, another
method must be employed. The same box, open end at the
bottom this time, or better still a covered brood box with a
few already-drawn combs in it, is placed *just above* the swarm
and secured in this position where it may need to remain for
a day or two. Usually it will happen that the bees, glad of a
ready-made home so close at hand, will climb up and into it.

Once they are in, the box can be removed, and if a brood
box has been used, it can be either united directly with the
weak colony (see page 85) or placed on a stand and allowed
to become established as a colony in its own right.

a brood box is placed just above the swarm

If any old box has been used, the bees must be housed in a proper brood box before they begin constructing comb and rearing brood. Wait until evening. Having prepared their new home with a few drawn-out combs and remaining frames fitted with foundation, place a board, covered with a white sheet, running up at a slight incline to the entrance. Carefully lift and carry the box of bees and hold it above the ramp. A sharp tap will dislodge the swarm from the box and they will fall onto the cloth. Immediately they will begin to move up the slope and into the hive. The keen observer should be able to spot the queen against the white background as she runs in.

Never attempt to run a swarm into a hive that is already occupied not even if it is the hive from which the swarm has issued earlier in the day, for at best the bees will pour out again quickly. At worst there will be the most awful carnage, as one colony defends its home against invasion.

Swarming is the natural way in which a colony of bees is propagated, and if you are wanting to expand the size of your apiary, a swarm will be most welcome. Generally however, swarming is found to be a nuisance since it weakens the parent colony. Consequently beekeepers have invented countless methods for trying to prevent and control the impulse.

First let it be said that there are many successful beekeepers who leave their bees pretty much to their own devices. Certainly they get plenty of swarms; but most will argue that they get plenty of honey, too. However, there is no doubt that if swarming can be prevented, yields will be much greater.

Among the many practices advocated are: regular weekly inspection of brood and destruction of all queen cells; provision of plenty of extra room in the hive; clipping of queens' wings; blocking the exit of the queen with patent "Swarm catchers"; selective breeding of non-swarming strains; removal and replacement of the queen, etc. etc. Reading all these different theories can leave one with a headache and the conviction to "let the buggers fly".

Before we try to find a solution to the problem of swarming we should ascertain why it is that a colony swarms in the first place. The most obvious reason is lack of space at the time when the amount of brood and nectar is expanding. Providing plenty of room, ahead of time, is certainly one form of prevention. Unfortunately, a colony so provided will often still swarm, just as crowded colonies will sometimes remain settled throughout the year. L.E. Snelgrove, whose observations and innovations are well-known to beekeepers, has drawn attention to an interesting situation that occurs in any healthy colony and may be taken as the true stimulus for swarming.

"As the spring advances the queen, under the stimuli of rising temperature and more liberal supplies of food presented to her by her attendants, gradually increases the rate of her laying until this reaches a peak or maximum. She may maintain this peak rate for a few days, after which her laying powers slowly decline. Since the time of emergence of the young bees from the cells lags behind that of the laying of the eggs by three weeks, it follows that a time comes when the bees are 'hatching' in increasing numbers while the queen is laying at a diminished rate. The result is that in the seasonal history of normal stock there is a period when large numbers of nurse bees, eager to feed larvae, and of wax secreters ready to build new comb can find little employment, and are therefore in excess of the hive requirements. This period coincides with the time when the foragers are ready to bring in quantities of fresh nectar and pollen. Instinctively the bees seem to realise that this is the favourable time for colonising; they quickly develop the swarming impulse and the beekeeper often finds it extremely difficult or impossible to prevent them from fulfilling their purpose" (note 9).

The idea that it is because the bees are underemployed that they develop the swarming impulse is borne out by the fact that an abundant honey flow, with all the work this provides for the bees, will usually delay swarming.

Mr Snelgrove goes on to suggest a method of prevention that is an ingenious as any and a lot more effective than most. A brief summary of his devilish plan, based on the assumption that "the presence of an excess of nurse bees induces the swarming impulse" so that "if by some simple means the nurse bees and field bees could be separated in the same hive, the impulse would not be developed" is included here out of interest, though anyone wishing to put it into practice should refer to the original text (note 9).

Two empty brood boxes are placed beside the hive on level ground or a board. The brood box of the hive is then opened and all the combs with brood and bees are placed in one empty box, while the combs without brood (again with bees) are placed in the other empty box along with the queen and a little unsealed brood. This box is then placed on the stand, and excluder put on top of it, a super above that and the box containing brood and bees is placed on top.

After three days a patent board is placed between the super and the top box, separating them by means of a perforated zinc screen and providing a new entrance to the top box, at the side of the hive. In this way the nurse bees caring for the brood are separated from the field bees and the queen, who continues laying as if nothing has happened. All the bees are happily occupied; there are no hold ups in honey production and no need to swarm.

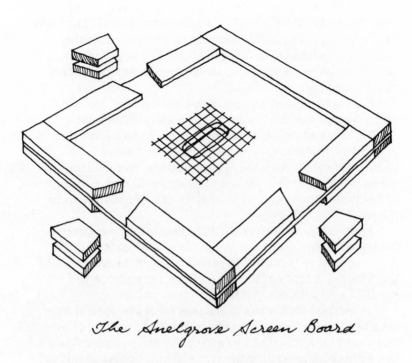

The Snelgrove Screen Board

As the brood hatches in the top box and the nurse bees take to the field, they are tricked into returning to the lower half of the hive by means of Snelgrove's cunning separating board, keeping the top box free of field bees and ensuring a maximum work force for filling the honey supers below. Any queens that may be reared in the top box, as a result of the absence of a queen there, will hatch, the first taking care to destroy the others. As there are no field bees there will be no swarm, the queen will mate and begin laying.

As the honey flow continues, supers are added *beneath* the separating board. At the end of the year the supers are removed; the perforated zinc taken from the middle of the separating board and the two colonies united. Thereupon the young queen will usually overcome the old one and the colony will settle in for winter strong in numbers and headed by a young queen — ideal conditions for a good start next year.

Of other more traditional methods, one in particular is not recommended: regular weekly inspection of the brood chamber and destruction of any queen cells found. The order of the hive is continually disrupted by this practice; also many people find the deliberate taking of life distasteful. Also this method is time-consuming, and by no means foolproof (note 10).

Clipping the queen's wings means that should she issue with a swarm she falls to the ground, and the bees not finding her, will return to the hive. Though swarming may be delayed by this method, nothing has been done to take care of the root cause. Also the assumption that because the queen is a small insect she does not suffer when a large part of her body is removed with a pair of scissors has yet to be proved.

Swarm-catching devices work by preventing the queen issuing with the swarm and are made of perforated metal sheeting, similar to a queen excluder. The main drawback to this method is that anything that traps a queen bee will inevitably trap drones as well.

One method that is worth considering if you intend to keep bees on quite a large scale, is selective breeding of a non-swarming strain. Queen rearing is a complex and fascinating subject quite outside the scope of this book and not to be undertaken by the beginner, though the possibilities for the commercial beekeeper are great: bees can be bred with greater resistance to disease, with milder temper, with larger bodies and even with longer tongues! To avoid the danger of random fertilisation, the queen being given an anaesthetic beforehand! However returning to the backyard . . .

Summer

Once all signs of swarming are past the beekeeper's main duty is to keep adding supers as they are required for the storage of honey. These should not be placed on top of the hive. Always first remove the existing topmost super; the new empty one goes in its place and the existing super is put back on top. This does not discourage the bees from filling the old super before they start work on the new, but will help to make them conscious of, and familiar with, the new space available. But most important of all this makes it easier to remove the full super at a later date without too much disruption of the colony. The whole procedure can then be repeated as space is required.

In the event of continued *very* hot weather, steps must be taken to provide extra ventilation by opening the top inner cover a little (not large enough for a bee to pass through). Also one super can be moved forward a fraction on top of another, providing a tiny gap at the back and front. Lack of ventilation can give rise to swarming and to melting and collapse of the comb. The beekeeper must be alert to any drop in the temperature and the cracks immediately closed.

There is not much more that can be done for the bees at this time. Only if the beekeeper has a garden or holding can he influence them by providing the type of plants that they most enjoy, flowers rich in pollen and nectar. Many of the bees' favourites will spring up whether you want them or not. If there is charlock around you won't need to provide any other flowers, as the bees will patronise this agricultural weed as long as it is in bloom. Dandelion too, though if you have an orchard you would be well advised to keep your dandelion population to a minimum as they yield about the same concentration of nectar as certain species of apple but are slightly more attractive to the bees, to the detriment of the cultivated plants.

Garden favourites are hollyhocks, wallflowers, forget-me-nots, and poppies, of which the Himalayan poppy is most favoured by bees. Aromatic herbs such as rosemary and thyme are always thick with bees when in flower and an unlikely shrub, cotoneaster, with its berry-like flowers is the scene of a perpetual buzz and hum during the spring and early summer. If you like wild flowers, try introducing mullein and evening primrose to your garden, both of which are useful medicinal herbs. Rosebay willowherb, with its many purple flowers abundant in nectar, is a good bee plant, though you would soon be overrun if you let this successful wild flower into your garden. After the Second World War, it sprang up all over London, finding bomb sites as hospitable as the forest fire clearings which it usually inhabits. City beekeepers enjoyed an unexpected increase in yields because of its appearance.

It is helpful for your bees if you grow a few early bulbs like crocus, hyacinth, snowdrop and tulip, so there is something around for them to visit on those rare days in February and March when the weather is mild enough for them to take wing.

A special mention should be made of borage, the herb with blue star-shaped flowers, that will grace any garden and keep the bees busy month after month.

Privet is about the only plant to avoid since its nectar will taint honey and make it virtually inedible, though there are certain species of lime, Tilia argentea and Tilia petiolaris, which are attractive and yet fatal to bees. Doubts have also been cast on ragwort, an unlikely enough introduction to anyone's garden, which, while it is said to taint honey, usually

94

flowers too late to have much effect.

The vegetable garden can also be a rich source of nectar. For instance the yellow flowers that develop from sprouting broccoli, if left in the ground until the seed pods form, will provide well for the bees, often at a time when the fruit blossoms are over but the clovers have not yet begun to flower. The Apis Club prints lists of all bee plants with much helpful information.

A garden stocked with well-chosen plants can thus be the source of twofold pleasure as you relax on a sunny afternoon enjoying the harmonious relationship that has existed between bees and flowers for millions of years, all the time aware of the honey surplus that swells day by day in your hive!

If the beekeeper notices that brood production is reduced or irregular he may decide that it would be desirable to have a new queen at the head of his colony.

Many methods of requeening are recommended, but the simplest is as follows: in the early spring the queen is allowed access to two brood boxes. At the start of the main honey flow these two boxes are divided. One box without the queen but with two or three combs of brood, is placed on a new stand. The majority of field bees will find their way back to the old site and so the original hive with the existing queen will not be much weakened, and the season will continue as normal. The bees in the newly-positioned hive will immediately start to rear a new queen.

Towards the end of the season the two hives are united, whereupon the young queen will depose the old, and requeening will have been deftly accomplished.

At the time when the main honey flow has virtually ceased, there is one major crop that is yet to be worked by the bees: heather. If there is a heather moor within a couple of hours run of your apiary, and you have suitable transport, the possibility of moving hives there just for the duration of the heather flowers is very interesting. The quantity of honey gained can be considerable, its flavour well known and unbeatable. Unfortunately, due to its very thick and crystalline consistency, it can be difficult to extract.

Autumn

Harvest time. When the honey flow is over, it is time to
remove the honey surplus. This is done with the aid of a
clearing board, a hive tool and usually a little smoke, a few
puffs of which should be directed into the hive entrance
before starting work. First remove all the supers and place
them at the side of the hive. The hive tool will be needed to
prise the boxes apart, for they will be firmly glued together
with propolis.

One of the advantages of the double walled hive is that the
outer lifts, having been first removed, can at this point be
used as stands for the supers.

Now it will be necessary to discover just how great the
honey surplus is, so an inspection must be made of the brood
chamber. To avoid having to feed sugar syrup to the bees
later in the year or the following spring, you will have to
ensure that they have fifty to sixty lbs of honey for their
own use. Either judge by the weight of the hive (a double
brood chamber with stands and covers, bees, comb and
sufficient honey should weigh about 1 cwt.) or by the amount

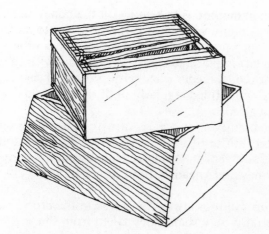

The outer 'lifts' can be used as stands for the 'supers'.

of capped honeycomb (not brood) that is visible. A shallow frame holds about 2lbs; a deep frame about 3lbs.

It is very unlikely that there will be this much honey in even a double brood chamber, so at least one super will have to be left in place. If this is the case, be sure to leave the queen excluder in position above the brood box. If there is sufficient honey without a super, the queen excluder can be removed.

Now place a clearing board, fitted with "bee escape", on top of the bees and their Winter store and then replace the remaining supers containing the honey surplus above the clearing board. The hive can now be reassembled and left for at least 24 hours, by which time the supers should be free of bees and can be taken to the "honey house" (see page 66).

Following this procedure you can be certain that any honey surplus you obtain will be from honey-fed bees. Unfortunately you cannot be certain to obtain much honey surplus! In a poor summer the bees may barely collect enough for their own needs. However, as time goes by, a carefully managed honey-fed colony will pay dividends in terms of its strength and health and the quality of honey produced.

A more common method is to remove all the honey supers and then feed the colony with sugar syrup or candy to make up their stocks. Once the supers have been taken away the clearing board is left in place above the brood box, with the 'bee escape' removed. A feeder filled with syrup is placed over the opening and refilled regularly until the required amount has been given.

Syrup is made by mixing two parts of white sugar with one part boiling water. Stir until all the crystals are dissolved and allow to cool before feeding. Candy is made by adding pure icing sugar to honey and kneading until firm, just as if making bread.

Since the syrup will thus always be stored in the brood box, one can be fairly sure that honey taken from the supers the following year will be pure: perhaps it is only the physiology of the bees that will have been weakened.

Just what you feed to your bees during the winter is up to you, though a barren summer followed by a fierce winter may force your hand and send you running to the arms of Messrs. Tate and Lyle.

If your bees have been working the heather the honey flow will not be over for you until the end of September, when any hives that have been moved to the moors should be brought home and, once the surplus has been removed, prepared for winter.

going to the moor

Before finally closing the hive for the winter a thorough inspection should be made in the same manner as in the spring (see page 85). Make a note of any parts that will need replacing next year. When positioning the cover, be certain to leave the proper "bee space" over the top of the frames, to enable the winter cluster to move freely and gain easy access to their store.

As the activity in the hive settles down to a minimum, the heaviest work of the beekeeper has only just begun. The supers, often weighing over thirty lbs apiece, have been carried to the honey house and then follows the lengthy task of uncapping and extracting.

The first of these jobs can be performed with any sharp, long-bladed knife, though there are special tools on the market to make work much easier. Simply, the cappings are sliced away from the comb, taking as little else with them as possible. The combs are then placed in an extractor (see page 66). Once extracted, the honey should be allowed to stand for a few hours in the sump of the extractor, or in a separate honey "ripener", before being bottled. All debris of wax particles and any air bubbles will rise to the surface and the resulting product will be clearer. Don't throw away the scum that collects on the surface: it is the "best part" and you will find that it is to the eating of these cappings that elderly bee-keepers attribute their long-lasting health and vigour.

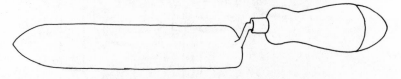

Uncapping Knife with 250mm blade

If honey is to be sold it should be bottled in 1 lb clear jars with a label stating the area of origin and the name and address of the beekeeper.

Honey for home consumption can be stored in larger vessels. It is good to mark each batch of honey, so that if you are able to build up any sort of reserve from one season to the next, you will have great fun comparing textures, flavours and colours. "An excellent year", "the sycamore was profuse that summer", "the wind was on the wrong side of the clover", etc. etc. The differences can be astonishing.

Once the combs have been extracted, they should be replaced in the supers and stored safely for use the following year.

If you have no access to an extractor, the combs can be cut from the frame (easier if unwired foundation was used – see page 72) and devoured in this whole state. The main drawback however is that you will be continually providing new foundation and the bees must continually be making new comb. This, however, in terms of swarm prevention, may not be a bad thing. Comb cut from the frames in this way can be broken up into smaller segments and bottled.

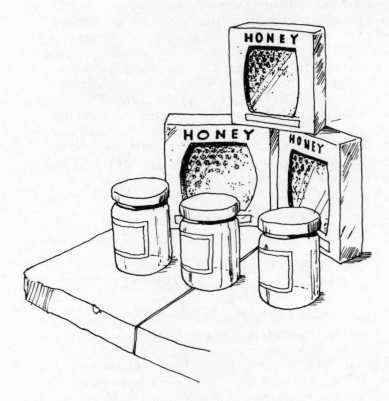

PART THREE:

Problems

Diseases of bees

ike all creatures living communally bees are subject to various diseases, some infectious and others due to some temporary environmental condition.

Diseases of the latter type, such as dysentry, can arise through improper diet, unfavourable weather, damp, starvation, or any accumulation of undue stress. A careful beekeeper should be able to avoid such situations with little trouble.

Infectious diseases can largely be prevented by cleanliness and care.

Worker Bee — Internal Organs

1 Brain	11 Dorsal Diaphragm
2 Salivary Orifice	12 Ventriculus
3 Sucking Pump	13 Heart
4 Salivary Duct	14 Mouth
5 Salivary Syringe	15 Ventral Diaphragm
6 Oesophagus	16 Tongue
7 Nerve Cord	17 Anterior Intestine
8 Honey Stomach	18 Rectum
9 Ostium	19 Anus.
10 Proventriculus	

Infectious diseases are more serious, harder to prevent, and often disastrous in their effect.

Acarine Disease. This is caused by a parasite that penetrates the breathing tubes of the bee and feeds on the bee's blood, finally multiplying so prodigously that the tubes become blocked and death ensues. If a hive is very listless and large numbers of older field bees remain inside on fine days, you may be in trouble, and should consult a local expert. County Bee Officers are always helpful and interested souls and can be contacted through the Local Government Offices.

Mosema Disease. This is caused by a smaller parasite that lives in the bee's stomach. It is not easy to detect and can prevail for long periods in a colony without much depleting the numbers, but is generally found to have a weakening influence. Queens are particularly susceptible to this disease.

Bee Paralysis. This can be identified by the spectacle of hairless, greasy individuals, trembling slightly with paralysed limbs, being dragged from the hive and cast upon the ground by other healthy members of the community. Not much is known about this disease.

Bees that have come into contact with toxic agricultural sprays may exhibit many of the same symptoms as above, but whereas infectious diseases can largely be prevented by cleanliness and care, the apiarist has little control over the activities of his "modern minded" neighbours.

European Foulbrood is, as the name suggests, a disease of the brood. It afflicts only weak colonies early in the year and can usually be eradicated simply by making the colonies strong.

American Foulbrood is on the other hand a much more serious disease and is best treated by destroying and burning the affected colony along with the frames, comb and all hive parts. The initial symptoms are discolouration of the brood which in later stages, as decomposition advances, becomes quite moist and evil-smelling.

The faintest suspicion of any such condition in your hive should prompt you to immediate action. Not only your colony but all those in the area are in peril.

Enemies of bees

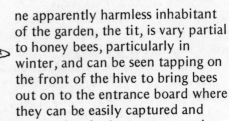

ne apparently harmless inhabitant of the garden, the tit, is vary partial to honey bees, particularly in winter, and can be seen tapping on the front of the hive to bring bees out on to the entrance board where they can be easily captured and devoured. The best remedy is a supply of other tasty morsels, peanuts or breadcrumbs, on a nearby bird table.

Mice will often enter occupied hives during the winter. The stress induced by such an intrusion can often by fatal to a weak colony. Another favourite nesting place of mice is amongst empty combs in stacks of stored supers. These will be quickly damaged, spoiled and unacceptable to bees.

Toads are as ready to make a meal of a bee as any other insect, though if the entrance to the hive is raised far enough from the ground, they will be unable to enter and will content themselves with feasting on the bodies of any dead bees on the ground in front of the hive. Considering the usefulness of toads in other areas of the garden, no attempt should be made to discourage their presence.

Ants and wasps are both honey lovers. Ants will happily enter a hive and enjoy the honey from uncapped cells. Wasps are more aggressive in their approach, intercepting heavily-laden bees and tearing them in half at the waist. They then carry off the abdomen, like a convenient bulk honey container, to their own nest for their own brood.

Bee lice (Braula coeca). This is a tiny insect which, though it attaches itself to the bodies of bees, is not parasitic but lives by coercing the workers to feed it with honey. The grubs of the bee lice will burrow through the capping of honey comb but leave the honey itself unharmed. Cleanliness within the hive is the best prevention.

The most common and real nuisance to beekeepers is the *wax moth*, the larvae which burrow through comb leaving behind criss-cross tunnels of desctruction strewn with pieces of silky web. A favourite spot for these grubs is in the comb of empty stored supers. Some sprigs of wormwood strewn amongst the frames is a good preventative, but should never be used in a hive occupied by bees.

In other climates, bees are surrounded by more unlikely dangers. In East Africa there is a species of bird called the honey bird which, upon discovering a colony of bees in a tree, will go in search of a certain beaver-like animal and guide it, by means of a call, to the honey hoard. The beaver opens up the tree and the two share the spoils. Local people have long since trained these birds to hunt honey for them, always being careful to leave a portion for the bird.

The Red-backed Shrike, or Butcher-bird, is so-named because it impales bees and other small creatures on Blackthorn spines, which it uses as a larder.

PART FOUR:

Success

Some products, honey, honey
vinegar, beeswax, section, mead.

The Harvest

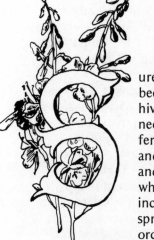

urely the most important product of the bees' labour is not to be found within the hive. In the course of their search for nectar and pollen they take care of the fertilisation of many wild flowers and trees and the greater part of all cultivated food and fodder crops. It is surprising that, whilst fruit growers are at such pains to increase their yields by means of expensive sprays and fertilisers, one often passes vast orchards with not a bee hive in sight. At the time of writing, the neglected apple trees in our orchard, which serves also as an apiary, are so laden with fruit that the mere thought of picking and storing it all is exhausting!

Seed yields in fields of clover have been increased twenty-fold by the provision of nearby colonies. We are in fact, heavily dependant on the bees of the world for our nourishment.

The best known bee product is honey, though one word is scarcely adequate to describe the variety of substances which can be extracted from the comb: as different from the flowers from which they come. The colour of honeys ranges from the pure white of clover, through the greens of sycamore and lime, the pale gold of acacia and dandelion to the dark and pungent chestnut and buckwheat.

It is rare indeed to meet someone who doesn't like *any* type of honey, though the powerful flavours of the darker honeys are too much for some, and clear acacia too mild for others. The "local" honey gathered from the rich and varied vegetation of the English countryside and garden is always the most popular.

Only privet, some laurels and ragwort produce nectar that, whilst attractive and harmless to bees, is as disgusting as it is injurious to humans.

Honey must be one of the oldest foods of man, and apart from milk is the only substance that is intended by nature solely as a food. Unlike refined sugar, honey has a soothing effect and is easily assimilated. It can be taken by the most sensitive, even ulcerated digestive system, when no other food is possible. Its many curative and medicinal properties have long been appreciated. It is most useful as a household remedy for burns: quickly and liberally applied, blistering will be prevented and the pain soothed.

Honey will absorb moisture from anything that it comes into contact with and, since moisture is necessary to maintain life in all living organisms, any bacteria being treated with honey will loose vital amounts of water and die. For this reason it has been used throughout the ages as a preservative, antiseptic and embalmer: there are few germs that can survive more than a day in honey. Properly stored it will keep indefinitely. When the Tomb of Queen Tyi's parents was opened, a jar of honey was found, still partly liquid, its full aroma and taste maintained. It had been there for some three thousand three hundred years. Honey has been known to cure most stomach ailments, tuberculosis, baldness, anaemia, and if the particular pollen which causes hay fever can be located, a little honey gathered from that same plant can be taken for relief. It is said to be both a diuretic and an aid in water retention (not having to get up in the night); a soporific (sends you to sleep) as well as a stimulant (wakes you up); even an aphrodisiac . . .

If applied externally, honey has a nourishing and refining effect upon the skin.

A diet of milk and honey supplemented by a glass or two of orange juice each day has been found sufficient to keep a man in good health and activity indefinitely.

Throughout the literature and the scriptures of the world, from Rig Veda to the present day, honey has been praised and celebrated. It contains up to thirteen different kinds of sugars notably Levulose and Dextrose; up to twenty-five minerals; a small amount of protein, enzymes, albumen and vitamins, as well as some ingredient that no one has been able to locate or identify: we can analyse honey, we can produce a synthetic substance that looks and tastes the same, but when eaten it has few of the same beneficial effects.

Though like most foods it is best eaten raw, honey can be used in place of sugar for all cooking needs. In baking and wine-making it makes an effective starter for the yeast and will also add its own fragrance. Try using honey as a sweetening for tea and coffee. There is certainly no nicer way of enjoying honey than eating it spread thickly on wholemeal bread and butter and yet once you become used to preparing food with honey endless possibilities for new and imaginative dishes will unfold before you. A few simple recipes are included at the end of this chapter to get you started. Remember when using honey in place of sugar in cakes to add less liquid to compensate for the moisture of the honey: the end result will stay moist longer, by the way.

It can be used for preserving fruit and being sweeter than sugar, less will need to be used. Why not try this interesting method which makes use of honey's own preservative powers, and not only dispenses with the deadly white stuff, but also avoids cooking and the consequent damage to the life of the fruit. Place the prepared fruit in a shallow dish and cover with honey. Place the dish in your solar wax extractor (see page 68) and leave in full sunlight for a few hours until thick. Put straight into sterilised jars and cap, leaving as little space for air on top as possible.

Even sweeties can be made with honey. For centuries a blend of sesame seeds and honey has pleased many a sweet tooth in all parts of the world. A more traditional English confection can be prepared as follows: boil two measures of honey until it sets hard when dropped into cold water. Stir in half a measure of butter, a little salt and lemon or vanilla for flavour; cut this into squares and allow to cool; you may then have produced butterscotch.

Delicious salad dressings, cool summer drinks and warm winter beverages can all be made from honey and lemon juice. But the best-known honey drink is Mead. You will need about 4lbs honey for each gallon of water. This can then either be placed in a container and allowed to do its own thing or can be boiled and, when cooled to body temperature, a little yeast added. Plenty of more complicated recipes abound.

SACK

4 lb of honey
3 or 4 fennel roots
3 or 4 sprays of rue
1 dessertspoon citric acid
2 gallons water and yeast

This favourite Elizabethan drink ought to be experienced. After washing the roots and leaves boil them in water for 45 minutes. Then pour the liquid through a sieve and add the honey. Re-boil the mixture for two hours taking off any froth or scum which appears. Cool the liquid to 70 degrees F, then add yeast and yeast nutrient and place in a fermenting bottle with an air-lock. It may ferment slowly. Bottle and rack after the sack has cleared. It should be drinkable after a year. Be careful not to put too much fennel into the mixture.

ROSE HIP MEAD

4 lb of honey
4½ lb of rose hips
juice from 2 lemons
yeast nutrient

Boil the rose hips in a gallon of water for 5 to 10 minutes and then let them cool. When cool mash them and strain it through muslin. Now add the other ingedients and stir until the honey has dissolved. When warm to the touch add the yeast and ferment.

SPARKLING MEAD

3 lb honey
1 tablespoon citric acid
1 gallon of water
Wine yeast and nutrient

Boil the water for about a minute and let it cool to 120 degrees F. At the same time heat the honey to 120 degrees F. and then mix the two stirring well until the honey is dissolved.

Let the mixture cool to 70 degrees F and add the acid, wine yeast and nutrient. Fill up the fermentation jug or bottle and keep the excess liquid next to it in a warm place at 65 to 70 degrees F. Top up the jug or bottle if the fermentation froths out. When it slows down, top up and fit an air-lock. When the fermentation has stopped completely move it into a cold room and leave it for three or four weeks before siphoning off the mead into a clean bottle or jar. Bung the neck with a rubber bung or a waxed cork. After six months siphon the mead off again and add 2 oz of honey dissolved into ¼ pint of water which has been boiled and allowed to cool to 70 degrees F. Mix it thoroughly and bottle it in strong bottles with either screw fittings or wire/tie the corks down.

You must add a yeast nutrient in order to aid fermentation.

Some other ideas

HONEY CHEESE

1 pint of strained honey
6 egg yolks
4 egg whites
juice of 4 lemons
rind of 2 lemons
3 oz fresh butter.

Stir the mixture over very slow heat until it thickens. Put into warm jars and seal. Will keep for a year or more in cool, dry place. Use it as a filling for tarts and sandwiches.

HONEY VINEGAR

1 lb honey
5½ pints of water

Stir thoroughly and cover with two thicknesses of fine muslin. Keep it at a temperature of about 80 degrees F. After about 6 weeks, if it tastes alright, strain it into another container. Stir in 1/4 oz isinglass which has been allowed to dissolve in a little water and let it stand for 2 weeks. Bottle and cork well. New corks must be used. Ideally wooden receptacles should be used throughout in the making of this vinegar.

GRANOLA — toasted breakfast cereal

1 measure of oatflakes
1 measure of wheatflakes
¼ measure of each of the following
    sunflower seeds
    hazelnuts
    cashews
    coconut (dessicated)

Mix all together with ¼ measure of honey and similar amount of oil (sunflower seed oil is best) and toast in a medium oven until golden. Add half a cup of raisins when cool.

A similar idea for an energy-giving snack that can be easily carried on hikes and rambles can be made by mixing together equal measures of your favourite nuts, seeds and dried fruit, chopping them up and mushing them with enough honey to sweeten and bind them together. Shape into little bars and wrap in rice paper.

A grapefruit cut in half and dressed with honey makes a nice way to start the day.

ICE CREAM

Use full cream milk and add some vanilla essence, finely chopped nuts and honey. Freeze it, stirring vigorously every now and again and consume in large quantities.

HONEY PIE CHEESE CAKE

Make a pastry crust and line a pie dish. Bake this crust on its own for about 10 mins. Mix together 1½ lbs cottage cheese, about 8 tablespoons of honey, some cinnamon and 5 eggs. Pour this mixture into the partly baked pie crust and return to the oven until the centre is firm. Serve cold.

HONEY CAKE

3 cups self-raising flour (85% extraction is best)
1½ teaspoons mixed spice
¼ cup chopped nuts
¼ cup raisins
¼ cup mixed peel

1 cup honey
½ cup cold coffee
2 tablespoons sunflower oil (or any other)
½ cup applejuice
4 eggs

Mix it all together and pour into oiled baking pans and bake at 300° for 1 hour.

PUMPKIN PIE

Make a pie crust.
Take 2 cups pulped pumpkin
    1 cup full cream milk
    6 tablespoons at least of honey
    4 eggs
    2 teaspoons of mixed spice

Mix and pour into the pie crust. Bake at 450° for 10 mins. and then at 325° until set (about 40 mins). Serve with cream.

Pollen. Also called bee bread. A visit to your local health food store, where pollen is to be found gaily packaged and loudly proclaimed as an elixir, may or may not convince you of its nutritive and medicinal potency. Suffice to say that many people consider the addition of a little pollen to their diet essential for good health. Some sufferers from hay fever have found great relief by taking small doses of pollen when afflicted.

Wax. At one time all candles used in church services were made of beeswax and even today this bee-product cannot be improved upon, for not only do beeswax candles give out a delicate fragrance but they will also burn more steadily and for much longer than other types.

The best polish for wood, flooring or furniture is made by adding a little turpentine to melted beeswax, the proportion depending simply on whether you want a thick or thin paste. Other uses for beeswax include: waterproofing cloth; insulating electrical components; impregnating paper cartons and cardboard liners for jar lids; dental impressions and of course the manufacture of comb foundation.

Propolis. Otherwise known as bee glue, this aromatic resinous substance can be a real nuisance to the beekeeper, for some strains of bees will use it more lavishly than others, covering every gap and crevice thickly. It does however, have a major use outside the hive. Drug companies make a preparation from propolis for use as an antiseptic. In one experiment at a German hospital, out of about 60 cases treated with the propolis-based antiseptic, there was not a single infection, while similar operations performed using other antiseptics were less successful. This should recommend it for the medicine cupboard where it may be kept for use on all minor cuts and burns. The best solvent for propolis is alcohol.

Royal Jelly. This milky, strong-smelling jelly is another bee-product to be found for sale as a dietary aid with powers of rejuvenation and regeneration. Certainly its effects on the female bee larvae are almost magical, changing the potentially stunted undeveloped worker into a voluptuous, highly fertile queen! Presumably it has been found to have different effects on the human physiology.

Venom. Surprisingly the most demonstrably curative effects of all bee-products have been achieved with bee stings. Almost immediate relief from arthritis and rheumatism often follows direct application of the venom. Best results are obtained by holding the bee against the affected part until it stings. Bee venom preparations sold by some drug companies are apparently less effective. So if you decide to keep bees, while you are certain to be stung many times, you are almost as certain to be free from rheumatism all your life. (Note 11).

Apart from such useful long-term benefits from bee stings, there are shorter-term, less pleasant effects: a couple of remedies that will serve to ease the pain and reduce swelling should be mentioned. Firstly the method of removing the sting is crucial. To grab it between thumb and finger, as one would spontaneously want to do, is to be avoided at all costs: this simply squeezes every last drop of venom into the skin. It is best to lift or flick the sting away, using a thumbnail.

Honey has long had a reputation as an effective antidote to bee stings and should be smeared on the skin as soon as possible.

Another remedy, and by far the best, is the good old dock leaf. Take a large leaf, squeeze and roll it hard between the palms of the hands for about 1 minute until the juice starts to run free, and apply this to the sting. You will notice almost immediate relief and subsequent lack of reaction.

As the months go by stings become a matter of course and

you'll probably find that a kind of immunity builds up in your system, so the pain is slight and the reaction negligible. Occasionally matters can turn the other way and one can become increasingly allergic to bee stings. If this should happen, there are various courses open to you: consulting your doctor is one; finding an alternative hobby at the first opportunity is another.

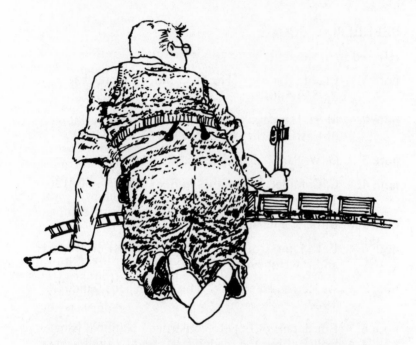

Books and Suppliers

BEEKEEPING BOOKS

referred to in the text:

note 1.　Claude Kellogg. "The Hive and the Honeybee". pp 557-566.

note 2.　M. H. Haydak. "Larval Food and the Development of Castes in the Honeybee" 1943.

note 3.　O. W. Park. "The Hive and the Honeybee".

note 4.　C. C. Miller. "Gleanings in Bee Culture". 30. 136. 1902.

note 5.　Butler. "Female Monarchy". 1609

note 6.　L. L. Langstroth. "A Practical Treatise on the Hive and the Honeybee". 1884.

note 7.　W. Herrod-Hempsall. "The Beekeeper's Guide". 1947.

note 8.　For details of how to construct a National (single-walled) hive, the reader is referred to a publication by New Age Access, entitlted "How to Build a Beehive". Write to New Age Access, P.O. Box No. 4, Hexham, Northumberland.

note 9.　L. E. Snelgrove. "Swarming — its Control and Prevention". 1934.

note 10.　ibid. p. 57.

note 11.　Dr. B. F. Beck. "Bee Venom Therapy". 1937.

for further reading:

"A.B.C. and X.Y.Z. of Bee Culture". A. I. Root. 1877
Frequently revised and reprinted since.

"The Complete Guide to Beekeeping". Roger A. Morse. 1973.

"The Dancing Bees: an account of the life and senses of the
honeybee". Karl von Frisch. Translation by Menthuen, 1970.

"British Bee Plants". Ed. A. F. Harwood. The Apis Club.

USEFUL ADDRESSES: Suppliers and Associations.

Robert Lee Ltd.,
Beehive Works,
George Street,
Uxbridge,
Middlesex, UB8 1SX.

E.H. Taylor Ltd.,
Beehive Works,
Welwyn,
Herts, AL6 0AZ.

Burtt and Son,
Stroud Road,
Gloucester.

R. Steele and Brodie,
Beehive Works,
Newport-on-Tay,
Fife, DD6 8PG.

E.H. Thorne Ltd.,
Beehive Works,
Wragby,
Lincoln, LN3 5LA.

British Beekeepers Association (over 100 years old;
publishers of "Bee Craft".)
55, Chipstead Lane,
Riverhead,
Sevenoaks,
Kent.

The B.B.K.A. should be able to put you in touch with your local beekeepers' association.

U.S.A.

The A.I. Root Co.,
Medina,
Ohio.

Dadant and Sons,
Hamilton,
Illinois.
(also publish the American Bee Journal)

Walter T. Kelley Co. Inc.,
Clarkson,
Kentucky, USA.

AUSTRALIA

John L. Guilfoyle Pty Ltd.,
Boundary Road,
Darra, Queensland 4076

Pender Bros Pty Ltd.,
P.O. Box 20,
Maitland, NSW 2320

Australian Bee Journal (monthly)
P.O. Box 137,
Noble Park, Victoria 3174.

Amateur Beekeeper's Association,
95 Waratah St.,
Brighton-le-Sands, 2216.

BACKYARD DAIRY BOOK

Andrew Singer and Len Street

This book is written as propoganda and its aim is to provide enough basic information to encourage readers to begin home dairy production. Thousands have reduced their dependence upon factory food by growing their own, or keeping chickens. Backyard Dairying is a further step towards self-sufficiency.

Contents
1. Why start?
2. Basic Economics
3. Which animal
4. Making a start
5. Feeding
6. Milking
7. A cow
8. Dairy products
9. A little technical information
10. Cream, butter, cheese and yoghurt
    References and Bibliography

"The chapters dealing with goat management are most convincing"—*Undercurrents*

"A very useful book"—*Sunday Times*

"It is remarkably clearly written and will be welcomed by many questing clients in a practitioner's waiting room"—*Veterinary Record*

BACKYARD POULTRY BOOK

Andrew Singer

The author of the Backyard Dairy Book brings the same treatment to the subject of domestic poultry keeping. This book will appeal both to the absolute beginner and the experienced backyarder keen to try out new ideas.

Contents
Stage One: Making the big Decision
    Chapter One: Why keep chickens?

Stage Two: Planning and Preparation
    Chapter Two: A little technical background
    Chapter Three: Which system to use?
    Chapter Four: Which hens to buy?
    Chapter Five: What to feed them?
    Chapter Six: Confining and housing them

Stage Three: Your Chicks Arrive
    Chapter Seven: Rearing chicks into layers
    Chapter Eight: When to re-stock?
    Chapter Nine: The harvest—eggs, meat etc.
    Chapter Ten: Diseases and problems

Stage Four: Other Poultry
    Chapter Eleven: Ducks
    Chapter Twleve: Geese
    Chapter Thirteen: Turkeys
    Chapter Fourteen: Guinea Fowl, Bantams and Pigeons

"Andrew Singer's comprehensive manual deals with all aspects of poultry keeping and is a must for all potential poultry keepers before making a start"—*Undercurrents*

PRACTICAL SOLAR HEATING
Kevin McCartney

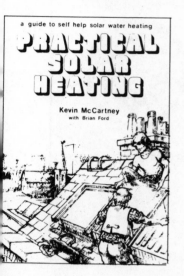

a guide to self help solar water heating

PRACTICAL SOLAR HEATING

Kevin McCartney
with Brian Ford

Even in a climate such as ours up to 50% of our domestic water heating could be supplied by the sun. So far only cost has postponed the widespread use of solar energy in homes throughout the country. Now, however, the rocketing price of conventional energy sources has made solar water heating economically competitive. Furthermore new plumbing materials and techniques now make do-it-yourself installation quite feasible, thereby greatly reducing the capital outlay.

This book has been written by a researcher at the Architectural Association who has designed, built and monitored numerous installations over the past five years. It is the most authoritative, thorough and inexpensive account yet to appear and many of the suggestions and precepts were tested during the successful installation of solar panels at Prism Press.

Contents

1. Solar Energy—what is it, why we should use it and how.

2. Basic Principles—Absorption, heat loss, greenhouse effect, heat and temperature, heat capacity and reaction time, tilt and orientation angles.

3. Solar Collectors—Function, types, surface finishes, materials, insulation casing and glazing.

4. Storage Tanks—Hot water cylinders, galvanised iron and plastic tanks, heat exchangers, cold feed tanks, expansion tanks, pressurised tanks and insulation.

5. Circulation—Thermosyphoning (gravity) and forced (pumped) circulation, pump and pipe sizes and materials, pipe lagging and controls.

6. Mounting the Collectors—Location, building permission, fixing over existing roof, removing roof and fixing collectors under glazing bars, wall collectors and free-standing collectors.

7. Choosing a System—Step-by-step guide to variations in methods of connecting collectors to storage tanks, frost protection and temperature boosters.

8. Plumbing for Solar Systems—Plumbing without a blowtorch, compression fittings, pipe types. With a blowtorch, capilliary fittings, low-cost plumbing.

9. Swimming Pools—Types of collector required, size, location and mounting, insulating the pool, effectiveness.

10. Examples—Commercial, Council and D.I.Y. installations.

11. D.I.Y. Collectors—Detailed plans for two types.

12. Installation Guide—Step-by-step instructions.

13. Survey of Manufactured Collectors.

KEEPING WARM FOR HALF THE COST
John Colesby & Phil Townsend

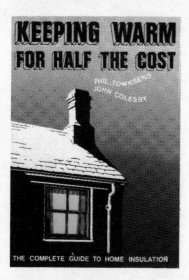

The complete guide to home insulation. The aim of this book is to show people how they can cut their heating bills in half and recover the initial cost within three years without going to extremes.

Contents
Why Insulate?
Basic Principles
Draughts
Roofs
Walls
Floors
Windows
Where Does All This Lead?

"To the best of our knowledge there are few such books about, and the authors have made a good job of presenting what could be a most involved textbook as a readable and concise layman's guide"—*Building*

Illustrated by over 100 line drawings
96 pages 8½" x 5½"
ISBN 0 904727 34 3 Hardback
ISBN 0 904727 35 1 Paperback

FOOD FOR THOUGHT
A Fresh Food Cook Book
Bill Scott & Marilyn King

"It would make a wonderful gift for anyone new to vegetarianism, or it would be greatly appreciated by the experienced cook, as so many of the recipes are really new and unusual, and even someone who hates cooking should find it delightful just to browse through and enjoy"—*Vegetarian*

"It is beautifully printed and designed with lovely illustrations and a lavish sprinkling of foody quotes from quite eclectic sources"—*Gay News*

"The co-authors, who mix the ingredients of imagination and charm with a nice dash of humour, give delicious, mouth-watering recipes from all over the world"—*Womans Journal*

156 pages 9" x 6"
ISBN 0 904727 20 3 Hardback
ISBN 0 904727 19 X Paperback